Mario F. Wullimann
Barbara Rupp and
Heinrich Reichert

Neuroanatomy of the Zebrafish Brain

A Topological Atlas

Birkhäuser Verlag
Basel · Boston · Berlin

Author(s)

Dr. M.F. Wullimann
Institut für Hirnforschung
Universität Bremen, FB 2
Postfach 33 04 40
D-28334 Bremen
Germany

Dr. B. Rupp
Zoologisches Institut
der Universität Basel
Rheinsprung 9
CH-4051 Basel
Switzerland

Prof. Dr. H. Reichert
Zoologisches Institut
der Universität Basel
Rheinsprung 9
CH-4051 Basel
Switzerland

A CIP catalogue record for this book is available from the Library of Congress, Washington D.C., USA

Deutsche Bibliothek Cataloging-in-Publication Data

Wullimann, Mario F.:
Neuroanatomy of the zebrafish brain : a topological atlas /
Mario F. Wullimann, Barbara Rupp and Heinrich Reichert. –
Basel ; Boston ; Berlin : Birkhäuser, 1996
  ISBN-13:978-3-0348-9852-2   e-ISBN-13:978-3-0348-8979-7
  DOI : 10.1007/978-3-0348-8979-7

NE: Rupp, Barbara:; Reichert, Heinrich:

Printed on acid-free paper produced from chlorine-free pulp. TCF ∞
Cover design: Micha Lotrovsky, Therwil

ISBN-13:978-3-0348-9852-2
9 8 7 6 5 4 3 2 1

Contents

Acknowledgments

Thanks are due to Carolin Pfau for preparing a histological whole head series of a zebrafish and to Ingrid Singh for advice and help with the photographical documentation. A number of colleagues have made important contributions and constructive comments: Drs. Peter Bartsch, Michael H. Hofmann, Catherine McCormick, Ulrike Niemann, Luis Puelles, Gerhard Roth, David G. Senn, Helmut Wicht. This work was supported by the Deutsche Forschungsgemeinschaft (Wu 211/1–2), the Freie Akademische Gesellschaft Basel, and the Swiss National Foundation.

1   Introduction: neuroanatomy for a neurogenetic model system

«Zebrafish hit the big time»: This title of a recent commentary in *Science* (Kahn, 1994) reflects the growing importance of a new model organism for developmental research, the zebrafish *Danio rerio* (Fig. 1). Pioneering work using this model began in the laboratories of George Streisinger and Charles Kimmel in the early 1970s, and to date studies of zebrafish have contributed notably to understanding early nervous system development and its genetic regulation in at least three areas:

1. Differentiation of first neurons and their associated tracts and commissures in the embryonic zebrafish brain
2. Neuromeres and expression of regulatory genes in embryonic zebrafish
3. Generation of zebrafish mutants via saturation mutagenesis

In this introduction, we will discuss briefly the relevance of a zebrafish brain atlas to these areas of research. However, rather than review each of these areas in detail, we will focus on how the neuroanatomy of the adult zebrafish brain might be helpful for future research in this model system.

Differentiation of first neurons and their associated tracts and commissures in the embryonic zebrafish brain

The first embryonic neurons in the zebrafish central nervous system (CNS) can be visualized immediately prior to final mitosis during neuromere formation and are found in a serially repeated sequence in the spinal cord (Hanneman and Westerfield, 1989; Bernhardt et al., 1990; Kuwada et al., 1990) and in the brain (Hanneman et al., 1988; Wilson et al., 1990). Soon thereafter, these neurons extend pioneering axons in a highly predictable, stereotyped fashion, and by 24 hours they form an early scaffold of tracts and commissures (Fig. 2; Metcalfe et al., 1986; 1990; Chitnis and Kuwada, 1990; Wilson et al., 1990; Wilson and Easter, 1991; Ross et al., 1992). This early axon scaffold includes eight tracts and four commissures (Fig. 2). At 48 hours, the number of fibers in these early tracts has increased by a hundredfold, and three more tracts and their commissures (intertectal and habenular commissures, commissure of the posterior tuberculum) have formed (Wilson et al., 1990), resulting in a total of eleven tracts and seven commissures. (Since many of the designations for early tracts

Figure 1. Adult zebrafish

and commissures only apply to the embryonic zebrafish brain, they are not listed in the index.)

The fact that many follower axons between 24 and 48 hours of development grow along the few axons constituting the early scaffold indicates that the first (pioneer) axons might provide cues to guide the follower axons (Wilson et al., 1990). However, the pioneer axons themselves and all axons approaching axonal intersections necessarily need additional guidance cues (Chitnis and Kuwada, 1991; Chitnis et al., 1992), indicating that multiple guidance cues may be required generally for orderly axonal outgrowth.

Many early regulatory genes have expression boundaries at positions where the neurons and tracts of the early scaffold are located, and this early morphological patterning can be changed systematically by experimentally altering gene expression (Wilson et al., 1993; Macdonald

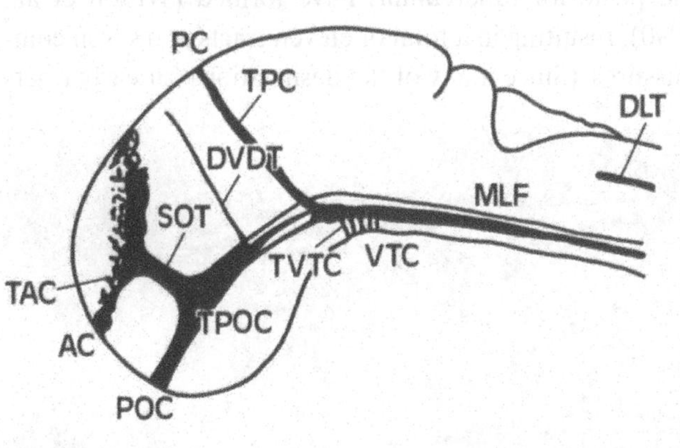

Figure 2.
At day 1 postfertilization, a simple scaffold of eight tracts can be visualized in the zebrafish immunohistochemically (see text). Four tracts are associated with commissures (tract of the anterior commissure (TAC), tract of the postoptic commissure (TPOC), tract of the posterior commissure (TPC), tract of the ventral tegmental commissure (TVTC). The four additional tracts are the supraoptic tract (SOT), the dorsoventral diencephalic tract (DVDT), the medial longitudinal fascicle (MLF), and the dorsal longitudinal tract (DLT). Also present in the telencephalon at day 1 (but not shown) are fibers of the olfactory nerve. Compiled from Metcalfe et al. 1986 & 1990, Chitnis & Kuwada 1990, Wilson et al. 1990, and Ross et al. 1992.

et al., 1994). This indicates that the early scaffold in the brain might reflect early regionalization processes. Thus, another hypothesis, namely, that the early scaffold is a transitory structure during embryogenesis, must also be considered. Documentation of the postembryonic transformation of embryonic CNS structures into their adult configuration in the zebrafish is needed in order to decide whether the early scaffold develops directly into adult nuclei and tracts or, alternatively, represents a transitory embryonic structure.

A striking similarity in the organization of the embryonic and adult zebrafish brain is that each of the seven commissures present at 48 hours corresponds topologically to adult commissures. They bear the same names in both life stages, except for the embryonic ventral tegmental commissure, which is called the ansulate commissure in adult teleosts. However, since we observe 13 adult commissures, the development of 6 commissures remains undocumented at present. The good topological correspondence between embryonic and at least some adult tracts and commissures does seem to suggest that they develop into each other by addition of fibers, though the alternative hypothesis that all embryonic tracts may be transitory must be kept in mind. The fact that all embryonic tracts are in a subpial position, i.e. they are located at the periphery of the CNS, in contrast to most adult tracts, which lie deep in the CNS, lends support to the latter hypothesis. Another way of comparing embryonic and adult zebrafish brains is to look at neuronal connections in both life-history stages. The embryonic neuronal connections constituting the early axon scaffold are shown in Figure 3, and the adult connections are discussed in chapter 6. In the embryonic and the adult zebrafish brain, the medial longitudinal fascicle (MLF) carries descending axons of neurons in the nucleus of the medial longitudinal fascicle (NMLF). Similarly, the embryonic dorsal longitudinal tract (DLT) carries descending primary sensory fibers of trigeminal ganglion cells, as does the adult descending trigeminal root (DV). Such cases support the hypothesis that parts of the early scaffold develop directly into adult structures. Other embryonic connections clearly are transitory, such as the ascending axons of the spinal Rohon-Beard cells in the DLT (Metcalfe et al., 1990) or the tegmental neurons, whose axons ascend in the tract of the postoptic commissure to cross in that commissure (Wilson et al., 1990); both connections are absent in the adult brain.

It is evident that a large gap of knowledge exists between the detailed documentation of neural development in the embryonic zebrafish and its adult neuroanatomy. Thus, the postembryonic development of the zebrafish brain must be studied before we can understand by which mechanisms the developmental transformation from the relatively simple embryonic condition to the complex adult brain occurs. As exemplified above, the atlas of the adult zebrafish brain is a necessary tool for doing so.

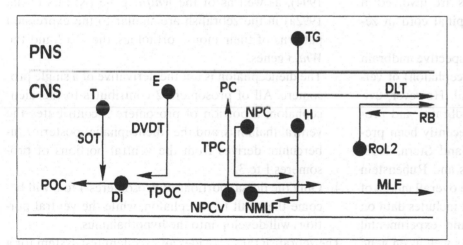

Figure 3.
Neuronal origins and pathfinding of axons that constitute the early scaffold of tracts shown in figure 2. Axons originate from clusters of primary neurons, which are shown here as single cells.
Abbreviations: CNS: central nervous system; Di: diencephalon; DLT: dorsal longitudinal fascicle; DVDT: dorsoventral diencephalic tract; E: epiphysis; MLF: medial longitudinal fascicle; NMLF: nucleus of the medial longitudinal fascicle; NPCv: ventral nucleus of the posterior commissure; PC: posterior commissure; PNS: peripheral nervous system; POC: postoptic commissure; RB: Rohon-Beard cells; RoL2: individual cell in the superior reticular formation; SOT: supraoptic tract; T: telencephalon; TG: trigeminal ganglion cells; TPC: tract of the posterior commissure; TPOC: tract of the postoptic commissure. Compiled from Metcalfe et al. 1986 & 1990, Chitnis & Kuwada 1990, Kimmel 1993, and Ross et al. 1992.

Neuromeres and expression of regulatory genes in the embryonic zebrafish

Recently, theories on the neuromeric organization of the vertebrate brain have enjoyed a renaissance. Although classical observations of repeated constrictions along the rostrocaudal axis of the vertebrate hindbrain were long disqualified as artifactual, today, the reality of neuromeres in the vertebrate hindbrain (rhombomeres) is substantiated by various approaches, such as the investigation of primary neurons, early axonal pathfinding processes, glial boundaries, proliferation zones, and, most important, differential gene expression coinciding with neuromere boundaries (Holland and Hogan, 1988; Keynes and Stern, 1988; Lumsden and Keynes, 1989; Murphy et al., 1989; Wilkinson et al., 1989a; 1989b; Lumsden, 1990; Wilkinson and Krumlauf, 1990; Clarke and Lumsden, 1993).

In terms of gene expression, rhombomeres are best characterized by different expression boundaries of homeobox genes of the *hox*-cluster and by the zinc-finger gene *Krox-20* (Holland and Hogan, 1988; Murphy et al., 1989; Wilkinson et al., 1989a; 1989b; Wilkinson and Krumlauf, 1990). In tetrapods, eight rhombomeres are recognized while only seven have been described for the zebrafish (Kimmel, 1993). Since rhombomeres 3 and 5 express the zinc-finger gene *Krox-20* in tetrapods (Wilkinson et al., 1989a; 1989b) as well as in the zebrafish (Oxtoby and Jowett, 1993), one must conclude that zebrafish lack the most caudal rhombomere. Also similar to tetrapods, the gene expression boundaries at the spinal cord-brain stem junction of two zebrafish *hox* genes (*hox-2.1* and *hox-2.2*) correspond well with those of their mouse orthologs (*Hox-2.1* and *Hox-2.2*), although in zebrafish they clearly extend into the most posterior hindbrain (Njølstad and Fjose,

1988; Njølstad et al., 1990). Another homeobox gene, the zebrafish *hlx-1* gene, is expressed in the hindbrain, where it reveals various rhombomeric and subrhombomeric boundaries in a temporal sequence in the zebrafish (Fjose et al., 1994).

Several genes of the homeobox-containing *engrailed* (*Eng*) gene family (Njølstad and Fjose, 1988; Hatta et al., 1991; Fjose et al., 1992; Ekker et al., 1992), one paired box-containing gene, *pax[b]* (Krauss et al., 1991a; 1991b; 1992b; Mikkola et al., 1992; Püschel et al., 1992b), and two genes of the *wnt*-family (*wnt-1* and *wnt[d]*; Molven et al., 1991; Krauss et al., 1992a) are expressed at and probably involved in forming the midbrain-hindbrain boundary in zebrafish. Moreover, the *wnt-1* (Molven et al., 1991), *wnt[a]* (Krauss et al., 1992a) and *pax[a]* (Krauss et al., 1991b; Püschel et al., 1992a) genes are involved in specifying the dorsal hindbrain and spinal cord in zebrafish.

The existence of neuromeres in the prospective midbrain and forebrain (diencephalon and telencephalon) of vertebrates is still somewhat controversial. However, evidence for mesomeres (prospective midbrain) and prosomeres (prospective forebrain) has recently been presented (Bulfone et al., 1993; Figdor and Stern, 1993; Puelles and Rubenstein, 1993). Puelles and Rubenstein (1993) present a predictive and testable overall model of vertebrate forebrain segmentation that includes data on descriptive (histology, morphology) and experimental (histochemistry, grafting) embryology as well as on gene expression data. This neuromeric model suggests that in addition to eight hindbrain neuromeres and one neuromere for the mesencephalon (mesomere), the forebrain consists of six more neuromeres (prosomeres). What is traditionally considered the diencephalic vesicle displays two segments, one for the pretectum (synencephalon, P1) and one that includes the adult epithalamus and dorsal thalamus (P2). Rostral to these two segments are four more prosomeres (P3–P6) constituting the secondary prosencephalon. Several points are important to note with respect to this neuromeric model:

1. As in the rhombencephalon, different gene-expression boundaries in the prosencephalon are co-localized with prosomere boundaries in the rostrocaudal axis. In addition, some prosencephalic gene-expression patterns coincide with boundaries in the dorsoventral plane of the brain.

2. Several expression patterns of early regulatory genes rostral to the rhombomeres in the zebrafish correspond well with the neuromeric model of Puelles and Rubenstein (1993). One of the *pax* genes (*pax[a]*) is expressed in alar plate regions of the future diencephalon and telencephalon (Püschel et al., 1992a), as is its mouse ortholog, the *pax-6* gene. Furthermore, the forebrain expression patterns of one homeobox gene of the *distal-less* family, *dlx2* (Akimenko et al., 1994), as well as of the *wnt[a]* gene (Krauss et al., 1992a) in the zebrafish are similar to the expression patterns of their mouse orthologs, the *Dlx2* and the *Wnt-3* genes.

3. The diencephalon is not the derivative of a single prosomere. All of prosomere 2 contributes to the diencephalon, a portion of prosomere 3 contributes the ventral thalamus, and the (diencephalic) posterior tuberculum derives from the ventral portions of prosomeres 1 to 3.

4. Only the dorsal portions of prosomeres 4 to 6 will become the adult telencephalon, while the ventral portions will develop into the hypothalamus.

The zebrafish thus provides an excellent test system for a neuromeric model of brain organization. A case in point is the posterior tuberculum, which is rather small in tetrapods (e.g. the inconspicuous mammalian subthalamic nucleus), but very large in teleosts (see chapter 6). Its large size makes it easier to study in zebrafish, and combined gene expression and grafting experiments are likely to reveal its embryonic origin, especially now that the neuroanatomy has been established.

There is a long way to go before the complex cascade of interactions of genes and their products in early vertebrate development is understood. The fact that various gene-expression patterns in the zebrafish correspond well with the neuromeric model of Puelles and Rubenstein (1993) is very encouraging for further research in this model system, which holds the promise of revealing fundamental neurogenetic patterns and mechanisms in vertebrates.

Generation of zebrafish mutants via saturation mutagenesis

The goal of saturation mutagenesis is to identify most – if not all – genes affecting early animal development and to understand their hierarchical, regulative role during development (for reviews, see: Mullins and Nüsslein-Volhard, 1993; Solnica-Krezel et al., 1994). This approach has recently been applied to the zebrafish system, and the results are impressive. Roughly 1300 zebrafish mutants have been described phenotypically by the Nüsslein-Volhard laboratory, and a further 600 were identified by the Driever laboratory (Kahn, 1994). The identification and cloning of mutant genes is now feasible: an initial genome map of the zebrafish has been constructed, and some of the loci identified by mutation have already been mapped out (Postlethwait et al., 1994). These advances in zebrafish research will greatly facilitate the investigation of the hierarchical interactions of regulatory genes during early ontogeny and, ultimately, may lead to an understanding of pattern formation relevant to vertebrate development in general.

Zebrafish mutants include not only abnormal phenotypes in very early development during the formation of germ layers and those showing massive body plan aberrations but also minor defects that become phenotypically only visible in later development. At the level of a detailed screening of CNS mutants, the present atlas may turn out to be useful, since many phenotypic defects in mutants are likely to differ from wild-type organization at the neuroanatomical level. In order to assess such defects, an atlas of the adult zebrafish CNS appears to be indispensable.

The zebrafish belongs to the teleostean family of cyprinids (minnows), which includes other neurobiologically well investigated species such as goldfish (*Carassius auratus*) and carp (e.g. *Cyprinus carpio*).

In a recent taxonomic revision (Barman, 1991), all species of the genus *Brachydanio* were reassigned to the genus *Danio*. Thus, the current species designation of the zebrafish is *Danio* (formerly *Brachydanio*) *rerio* (Hamilton-Buchanan, 1822). As a cyprinid, the zebrafish belongs to the large group of ostariophysan teleosts (6000 species, Lauder and Liem, 1983). Most ostariophysans are characterized by a sophisticated apparatus – the Weberian ossicles – to transmit sound from the swim bladder to the inner ear. The ostariophysans belong to the most derived group of teleosts, the euteleosts (17 000 species), which have three more basic teleostean outgroups, the osteoglossomorphs, elopomorphs, and clupeomorphs (1000–2000 species).

Zebrafish live in large schools and are communal breeders. Their natural habitats are the freshwaters of South and East Asia, i.e. India, Pakistan, Bangladesh, Burma, Sri Lanka, and Nepal (Barman, 1991).

3 Technical details

Histology

Fifty adult specimens of *Danio rerio* were processed in the course of this study. The animals were deeply anesthetized in methane sulfonate salt (Sigma, Deisenhofen, Germany) before being transcardially perfused with cold 0.1 M phosphate buffer (Sörensen; pH 7.4) followed by cold AFA (90 ml: 80% ethanol, 5 ml: 37% formalin, 5 ml: glacial acetic acid). The fish were then decapitated, and the skulls were opened dorsally to expose the brains. After one day in fixative, the brains were removed from the skulls and again postfixed for at least another month prior to further processing. The brains were then dehydrated, embedded in paraffin, and cut at 12 μm into either transverse, sagittal, or horizontal series of sections. To determine gross anatomy, some brains were fixed as described, removed from the skulls, and investigated under a stereomicroscope.

One specimen of *Danio rerio* was processed similarly, but the whole head was fixed for 15 days after perfusion and then decalcified in Titriplex III (Merck, Darmstadt, Germany) for 19 days. Subsequently, the head was put in Na_2SO_4 overnight and then rinsed in tap water, again overnight, before dehydration and embedding in paraffin. This specimen was cut at 10 μm into a transverse whole head series.

All sectioned brains were stained according to the procedure of Bodian, which reveals neurofilaments (see: Romeis, 1989), and counterstained with the Nissl-stain cresyl violet.

Preparation of figures

One series each of the transverse, sagittal, and horizontal histological sections was chosen for photographic documentation. While the numbers given in the atlas figures depicting the transverse plane represent the actual section numbers, the most lateral and dorsal sections shown in the sagittal and horizontal planes start arbitrarily with section number 1. However, the differences between figure numbers in the sagittal and horizontal series also represent the actual distances between sections.

The whole head transverse series served as a reference for structures that are almost always torn apart by removing the brain from the skull. The pituitary shown in transverse sections was photographed in this series and added to the photographed transverse sections with the aid of computer reconstruction. This was not done for the epiphysis, since it is well demonstrated in its full extent in some of the sagittal sections. Atlas photographs and corresponding ink drawings were computerized for further processing. Using computer techniques, artifacts and other undesired structures such as meninges or ventricular contents were removed, and graphics were added (e.g. background shading and lettering).

4 The brain of the zebrafish *Danio rerio*: an overview

The following account outlines the major CNS divisions in the zebrafish and the organization of these divisions into nuclei or laminae, including a description of major tracts and commissures. CNS divisions will be dealt with according to the classical anatomical sequence: telencephalon, diencephalon, mesencephalon, metencephalon (including cerebellum), myelencephalon, and medulla spinalis. Since many tracts and commissures caudal to the diencephalon extend into several brain parts and even into the spinal cord, they are treated in a final separate section. When appropriate, discrepancies between the neuromeric model of Puelles and Rubenstein (1993) outlined above (see: Introduction) and classical brain divisions will be discussed.

The principal terminology applied to nuclei and larger CNS divisions is indicated at the beginning of each major section. For general review articles on fish neuroanatomy, the reader is referred to Nieuwenhuys (1963), Northcutt and Braford (1980), Northcutt and Davis (1983), Nieuwenhuys and Pouwels (1983), and Nieuwenhuys and Meek (1990).

Tract designations often imply a specific origin and target of their axons. The present text includes only nomenclature on tracts that have been confirmed with experimental neuronal tracing. However, this does not imply that the interconnection indicated in the tract designation as the dominant one is the only one running in a specific tract. Unless unambiguous connectional information already exists in the zebrafish or other teleosts, we remained conservative in using tract designations that imply a specific connection, to allow future terminology to develop meaningfully as refinements of connectivity in the zebrafish brain emerge.

The relative proportions of the major brain divisions of the zebrafish (Fig. 4) reveal some features typical of cyprinids, i.e. relatively large vagal and facial lobes, although these are not as pronounced as in other cyprinid species (such as goldfish or carp). Further comments pertaining directly to specific brain variability in teleosts are given in the appropriate paragraphs of the present chapter, which describes the zebrafish brain. A more general discussion of the functional anatomy of the teleostean brain follows in chapter 6.

Telencephalon

The terminology of Nieuwenhuys (1963) as modified by Northcutt and Davis (1983) is applied except where noted. In teleosts, the topology of the telencephalon (Tel) is highly distorted (Nieuwenhuys and Meek, 1990). In all other vertebrate groups, the telencephalic hemispheres develop by paired evagination and thickening of the most rostral embryonic neural tube, and each hemisphere contains a lateral diverticulum of the ventricle centrally. In ray-finned fish, and most pronounced in teleosts (including the zebrafish), the roof plate of the embryonic telencephalon extends laterally with the effect that the paired alar plates forming the hemispheric walls roll out lateroventrally, a process called eversion. Thus, it is very difficult to infer from the adult topology of teleostean telencephalic cell masses alone their homologous structures in other vertebrates, though some advances have been made recently (see chapter 6).

The most rostral telencephalic divisions are the paired olfactory bulbs. The primary olfactory fibers (nervus olfactorius, I) entering the olfactory bulbs are the axons of the olfactory receptors, which are of placodal origin and, by definition, not part of the CNS. The rest of the telencephalon comprises two subdivisions, area dorsalis and area ventralis telencephali.

Olfactory bulbs

The olfactory bulbs (OB) form paired rostroventral protrusions adjacent to the remaining telencephalon and are interconnected with the latter by two tracts carrying secondary olfactory fibers: the lateral and the medial olfactory tracts. The lateral olfactory tract (LOT) enters the dorsal area of the telencephalon more rostrally compared with the more caudal entrance of the medial olfactory tract (MOT) into the ventral area of the telencephalon. Each olfactory bulb consists of four laminae that are roughly concentrically arranged throughout most of the bulb's extent, starting peripherally with a primary olfactory fiber layer (POF, present only in the rostroventral olfactory bulbs), followed by a glomerular layer (GL), an external cellular layer (ECL, including the efferent large mitral cells whose axons form most of the lateral and medial olfactory tracts), and an internal cellular layer (ICL). A distinct secondary olfactory fiber layer present in many other teleosts rostral to the formation of the lateral and medial olfactory tracts between the internal and external cellular layers is not apparent in the zebrafish.

Area ventralis telencephali
The ventral telencephalic area (V) is divided into periventricularly located nuclei which are in contact with the median ependymal lining of the ventricle, i.e. dorsal, ventral, supracommissural, and postcommissural nuclei (Vd/Vv/Vs/Vp), and nuclei which have migrated away from the ependyma to various degrees, i.e. central and lateral nuclei, dorsal and ventral entopeduncular nuclei (Vc/Vl/ENd/ENv). Periventricularly, a dorsal (Vd) and a ventral (Vv) nucleus lie rostral to the anterior commissure. The dorsal nucleus (Vd) is contiguous caudally with the supracommissural nucleus (Vs), which in turn is contiguous with the postcommissural nucleus (Vp). A periventricular nucleus called Vn (another nucleus of Nieuwenhuys, 1963) in some other teleosts is not present in the area ventralis telencephali of the zebrafish.
In the rostral, precommissural telencephalon, the migrated nuclei of the area ventralis include the central (= commissural) nucleus (Vc) and the lateral nucleus (Vl). While the central nucleus is located immediately lateral to the dorsal nucleus (Vd), the lateral nucleus has migrated to the very periphery of the brain. A cell-sparse neuropilar region is present between Vl and Vd/Vv. A distinct intermediate nucleus of area ventralis, described in some other teleosts as a lateral extension of the postcommissural nucleus, is absent in the zebrafish.

At caudal telencephalic levels, the lateral nucleus of area ventralis is replaced by the entopeduncular cell clusters. These form separate ventral (ENv) and dorsal (ENd) bands of tightly packed cells starting at the level of the anterior commissure. While the dorsal band of entopeduncular cells disappears more rostrally, the ventral band located around the lateral forebrain bundle extends into the telencephalo-diencephalic boundary (= preoptic) region. The dorsal entopeduncular nucleus may be homologous to the intermediate nucleus of the area ventralis of other species, although no close morphological association with the postcommissural nucleus, which is characteristic for other teleosts, is seen in the zebrafish.

Area dorsalis telencephali
Dorsal to the area ventralis telencephali lie the large cell masses of the area dorsalis telencephali (D). Since most of area dorsalis is developmentally everted, the ependymal lining of the unpaired median ventricle (TelV) continues dorsal to Vd into the medial zone of the dorsal telencephalic area (Dm) and becomes located at the apparent periphery of the telencephalon. A thin tela chorioidea covering most of D encloses the paired telencephalic ventricles. However, this tela has been removed in all our preparations.
Through most of the rostrocaudal extent of area dorsalis, four distinct cell masses, the medial (Dm), dorsal (Dd), lateral (Dl), and posterior (Dp) zones, surround a central zone (Dc). The sulcus ypsiloniformis (SY; Nieuwenhuys, 1959) separates Dd from Dm. Nucleus taeniae (NT) is located immediately ventral to Dp. In the most rostral as well as in the most caudal parts of the zebrafish telencephalon, medial and lateral zones of area dorsalis merge without an apparent boundary.
Although medial, central, and lateral zones of area dorsalis in the zebrafish are largecell aggregates extending over most of the telencephalon, we chose to await first connectional and immunohistochemical data in this species before further subdividing these areas, as has been done for *Salmo* and *Lepomis* (Northcutt and Davis, 1983). These

species are only distantly related to the zebrafish, and their telencephalic organization is sufficiently different to make a direct comparison of detailed subdivisions difficult at present.

Telencephalic tracts and commissures
The anterior commissure is divided into a dorsal part (Cantd) lying immediately ventral to the supracommissural nucleus of area ventralis telencephali, and a ventral part (Cantv) which abuts the preoptic region dorsally. In the precommissural telencephalon, several fascicles converge rostrocaudally and eventually fuse to form the lateral forebrain bundle (LFB). This is a compact tract located lateral to the medial forebrain bundle (MFB) and both tracts extend into the diencephalon.

The medial olfactory tract (MOT) enters the area ventralis telencephali ventrally and courses caudally, lateral to the ventral nucleus, to the level of the anterior commissure. The lateral olfactory tract (LOT) enters the area dorsalis telencephali more rostrally and courses laterally to the ventral border of the posterior zone of area dorsalis telencephali.

Diencephalon (including synencephalon and pretectum)

The terminology of Braford and Northcutt (1983), Northcutt and Wullimann (1988), and Wullimann and Meyer (1990) is applied except where noted.

The diencephalon proper has five major divisions which, in the adult brain, appear in a dorsoventral arrangement. They are the epithalamus, dorsal thalamus, ventral thalamus, posterior tuberculum and hypothalamus. The area praeoptica, although often considered part of the hypothalamus, is treated here in its own right because it constitutes an intermediate region between telencephalon and diencephalon. The synencephalon (the intermediate region between the dorsal diencephalon and mesencephalon) and the pretectum will also be treated in this section because these regions are intricately intermingled with diencephalic cell groups. The different meaning of

synencephalon in the neuromeric model of Puelles and Rubenstein (1993) is discussed below.

Area praeoptica
The preoptic area encloses the most rostral diencephalic ventricle (DiV) and can be divided into a ventral series of parvocellular preoptic nuclei and a dorsal series of magnocellular preoptic nuclei. The anterior parvocellular preoptic nucleus (PPa) extends rostrally to become located ventral to the telencephalon. It is replaced in the diencephalon by the posterior parvocellular preoptic nucleus (PPp). The suprachiasmatic nucleus (SC) is found ventrolateral to the latter.

Dorsal to the posterior parvocellular preoptic nucleus a magnocellular preoptic nucleus (PM) emerges, which is followed more caudally by the gigantocellular part of magnocellular preoptic cells (PMg). A parvocellular part of the magnocellular preoptic nucleus, as described in the goldfish (Braford and Northcutt, 1983), is not evident morphologically in the zebrafish.

Epithalamus
The epithalamus consists of the dorsal (Had) and ventral (Hav) nuclei of the habenula, and two dorsal outgrowths, the epiphysis (E = pineal organ) and the saccus dorsalis (SD). The latter is a chorioideal tela, whereas the former is an endocrine and light-sensitive organ.

Dorsal thalamus
The dorsal thalamus (DT) emerges below the caudal portion of the ventral habenular nucleus and consists of an anterior nucleus (A) and two more caudally situated nuclei, the dorsal posterior thalamic (DP) and central posterior thalamic (CP) nuclei.

Ventral thalamus
The rostral part of the ventral thalamus (VT) is located directly below the rostral part of the ventral habenular nucleus since it reaches far more rostrally than the dorsal thalamus does. The ventral thalamus is located ventral to

the dorsal thalamus only in its caudal extent. The ventral thalamus consists of an intermediate nucleus (I) as well as ventromedial (VM) and ventrolateral (VL) nuclei. The rostrolateral nucleus (R) of Butler and Saidel (1991) is also present in the zebrafish. It lies somewhat lateral to the ventral thalamic nuclei just described and is likely to belong to these nuclei. However, Anken and Rahmann (1995) recognize it as part of the entopeduncular complex.

Posterior tuberculum

As in other teleosts, the posterior tuberculum is much larger than both the dorsal and the ventral thalamus in the zebrafish. Its periventricular part consists of two nuclei, the periventricular nucleus of the posterior tuberculum (TPp) and the posterior tuberal nucleus (PTN), both of which are located between ventral thalamus and hypothalamus, as well as of the paraventricular organ (PVO). The migrated nuclei of the posterior tuberculum include primarily the preglomerular nuclei, which are – in rostrolateral to caudomedial sequence – the anterior, lateral, medial, and caudal preglomerular nuclei (PGa/PGl/PGm/PGc). The (preglomerular) tertiary gustatory nucleus (TGN), which has been misinterpreted as the visual nucleus glomerulosus of other teleosts (see chapter 6), is located ventral to the medial preglomerular nucleus. Ventrolateral to the caudal preglomerular nucleus lies the so-called corpus mamillare (CM), which is also included in the posterior tuberculum here. Additional nuclei which are likely to be part of the posterior tuberculum are the posterior thalamic nucleus (P), the nucleus subglomerulosus (SG), and the torus lateralis (TLa).

Hypothalamus

As in other teleosts, the hypothalamus of the zebrafish is by far the largest diencephalic area and includes ventral, dorsal, and caudal zones. While the ventral and caudal zones form most of the median tuberal portion of the hypothalamus (TH), the dorsal zone is additionally expanded laterally and includes the paired inferior (or lateral) lobes (IL) of the hypothalamus. These are separated from the tuberal hypothalamus by a deep ventral sulcus. All three hypothalamic zones display primarily periventricular cell masses which are flanked laterally by separate migrated nuclei.

The ventral hypothalamic zone with the periventricular nucleus (Hv) extends farthest rostrally. The anterior tuberal nucleus (ATN) and the lateral hypothalamic nucleus (LH) lie lateral to it. More caudally, the dorsal hypothalamic zone emerges and forms paired laterocaudal ventricular recesses (LR) surrounded by periventricular cell masses (Hd). Presumably migrated from these periventricular cell masses are the small-celled diffuse nucleus (DIL) and the larger-celled central nucleus (CIL) of the inferior lobe. The caudal protrusion of the inferior lobe lies ventral to the tegmentum and includes in its dorsomedial aspect the mammillary body (CM, which more likely belongs to the posterior tuberculum). The caudal hypothalamic zone starts out rostrally as a thick periventricular nucleus (Hv), which lies ventral to the posterior tuberal nucleus of the posterior tuberculum. More caudally, the caudal hypothalamus consists of paired posterior ventricular recesses (PR) surrounded by periventricular nuclei (Hc). The pituitary (Pit) is ventrally attached to the ventral and caudal hypothalamic zones. A saccus vasculosus seen in many other teleosts caudal to the pituitary is absent in the zebrafish.

Synencephalon

According to Braford and Northcutt (1983), the synencephalon designates a series of structures which are intermediate between the dorsal diencephalon and mesencephalon. These structures are located in the vicinity of the posterior commissure and include the nucleus of the medial longitudinal fascicle (NMLF), the periventricular pretectum (PPd/PPv), the paracommissural nucleus (PCN), and the subcommissural organ (SCO). Note, however, that in the neuromeric model of Puelles and Rubenstein (1993), the synencephalon is the most caudal prosomere (P1) and gives rise to all pretectal nuclei.

Pretectum

The pretectum is one of the most variable brain regions in teleosts (Northcutt and Wullimann, 1988; Wullimann and Meyer, 1990; Butler et al., 1991; Wullimann et al., 1991b). Following Braford and Northcutt (1983), we recognize a periventricular pretectum (see this chapter: Synencephalon), a central, and a superficial pretectum. The number and morphology of superficial and central pretectal nuclei in the zebrafish clearly represent the reduced pattern of pretectal organization in teleosts, in which nucleus corticalis as well as nucleus glomerulosus are absent (see chapter 6). The superficial pretectum is embedded in the optic tract and includes a parvocellular (PSp) and a magnocellular (PSm) nucleus. The PSm is caudally contiguous with the posterior pretectal nucleus (PO), whose cells appear to be slightly less large and less orderly arranged around a central neuropil compared with the magnocellular superficial pretectal nucleus. The central pretectal nucleus (CPN) is found dorsal to PSm and PO. The accessory pretectal nucleus (APN) lies dorsolateral to the posterior pretectal nucleus. Two additional nuclei are sometimes considered part of the pretectum: the dorsal accessory optic nucleus (DAO), which is ventrally adjacent to PSm and PO, and the ventral accessory optic nucleus (VAO), which lies more ventromedially. The VAO is very large in zebrafish compared with other cyprinids.

As mentioned already, all pretectal nuclei discussed here may derive from prosomere 1 and form the larger synencephalon of Puelles and Rubenstein (1993). Another hypothesis of interest here is that the superficial pretectal nuclei (PSp/PSm) may be homologous to the griseum tectale of birds (Marin and Puelles, 1994), in which case the superficial pretectal nuclei would be a derivative of the mesencephalic vesicle (mesomere) and not of prosomere 1 (L. Puelles, personal communication).

Diencephalic tracts and commissures

Starting in the telencephalon, both lateral and medial forebrain bundles (LFB/MFB) extend as distinct tracts into the diencephalon, the lateral one dorsal to the preglomerular area and the medial one approaching the posterior tuberal nucleus.

Ventral to the anterior preoptic region, the optic nerves (nervus opticus, ON, II) interdigitate in the optic chiasm (CO) and reach the contralateral side of the brain. Beyond the chiasm, the optic fibers are conventionally called the optic tract (OT), which is composed of a dorsomedial (DOT) and a ventrolateral (VOT) optic tract. The DOT runs dorsally along the lateral surface of the posterior preoptic and thalamic regions towards the optic tectum, while the VOT courses caudally along the ventral boundary of the optic tectum.

The pretecto-mammillary tract (TPM) originates in the magnocellular superficial pretectal nucleus and runs through the posterior pretectal nucleus to terminate in the mammillary body. The habenular nuclei project via the fasciculus retroflexus (FR; = tractus habenulo-interpeduncularis) to the interpeduncular nucleus (NIn; Villani et al., 1994).

The habenular commissure (Chab) runs between the two dorsal habenular nuclei.

Immediately caudal to the optic chiasm lies the large postoptic commissure (Cpop; = supraoptic commissure). The minor and transverse commissures are included here in the postoptic commissure, since they were not as clearly separable as in other species. Ventral to the postoptic commissure, the horizontal commissure (Chor) crosses the midline within the ventral hypothalamic zone. The tracts forming the horizontal commissure run caudally on each brain side, ventral to the preglomerular area, beyond which they turn first dorsally and then rostrally again towards the anterior tip of the optic tectum. All along their course, these tracts are also referred to as the horizontal commissure. A third diencephalic commissure lies caudal to the posterior tuberal nucleus. This is the commissure of the posterior tuberculum (Ctub; Herrick, 1948).

The posterior commissure (Cpost) is located in the region treated earlier as synencephalon.

Mesencephalon

The terminology of Nieuwenhuys and Pouwels (1983) is applied except where noted.

The mesencephalon includes, dorsally, the (multisensory) optic tectum and, ventrally, the torus semicircularis and the tegmentum.

Tectum opticum

The optic tectum (TeO) is the most complex layered structure in the zebrafish brain. It consists of four zones (periventricular grey zone, deep white zone, central zone, and superficial grey and white zone), which can be further subdivided into 15 layers (Northcutt, 1983). Different from all other vertebrates, the most superficial tectal layer in teleosts does not consist of retinal fibers (see chapter 6: Vision). This marginal layer consists of axons whose perikarya are in the torus longitudinalis (TL). The latter is a paired, longitudinal eminence of granular cells attached to the tectum. It is located in the medial tectal ventricle (TeV) and only occurs in ray-finned fish (for a review, see: Wullimann, 1994). The intertectal commissure (Ctec) runs between the tectal hemispheres.

Torus semicircularis

The sensory torus semicircularis (TS) is the mesencephalic target of ascending octavolateralis systems and lies on top of the lateral tegmentum from where it bulges out into the tectal ventricle. In cyprinids, the central nucleus (TSc) is related to audition and the ventrolateral nucleus (TSvl) is related to mechanoreception (Echteler, 1984; McCormick and Hernandez, 1996).

Tegmentum

The term tegmentum is used ambiguously in the literature. In mammals, the roof of the mesencephalon consists of the superior colliculus (tectum opticum of other vertebrates, which is part of the visual system) and the inferior colliculus (torus semicircularis of other vertebrates, which is part of the auditory system). The ventral mesencephalon is separated from this roof by the ventricle and forms the tegmentum, which has a dominant role in motor functions. The tegmentum arises embryonically from the basal plate, in contrast to the alar plate-derived, sensory-related tectum opticum and torus semicircularis.

The tegmentum includes many motor structures, such as the oculomotor (NIII) and trochlear (NIV) nerve nuclei, the parasympathetic Edinger-Westphal nucleus (EW), the nucleus ruber (NR), and the most rostral portion of the superior reticular formation (SRF). It also harbors the dorsal and rostral tegmental nuclei (DTN/RT), the perilemniscal nucleus (PL), the nucleus of the lateral lemniscus (NLL), and the interpeduncular nucleus (NIn). The rostral tegmental nucleus as defined by Grover and Sharma (1981) is homologous to the lateral thalamic nucleus defined by Braford and Northcutt (1983). However, we prefer Grover and Sharma's term here since they were the first to show that in cyprinids a projection to the optic tectum originates in the rostral tegmental nucleus. The nucleus of the lateral lemniscus is defined according to Prasada Rao et al. (1987), who showed that this nucleus projects to the spinal cord in goldfish. Recently, Becker et al. (1995) have confirmed this for zebrafish.

Whereas the axons of the oculomotor nerve (nervus oculomotorius, III) exit the brain ventrally between tegmentum and inferior lobe, the trochlear motor nucleus sends its axons dorsally, where they decussate (DIV) in the valvula cerebelli, turn caudolaterally, and exit the brain as the trochlear nerve (nervus trochlearis, IV) between torus semicircularis and rhombencephalon.

The tegmentum is bordered rostrally by the synencephalon, the dorsal thalamus, and the posterior tuberculum; ventrally by the hypothalamus; and dorsolaterally by the torus semicircularis. Caudally, the tegmentum is contiguous with the medulla oblongata without a clearcut morphological boundary. The interpeduncular nucleus and the trochlear nucleus are often considered to be the most caudal tegmental nuclei. However, the rhombencephalic griseum centrale as well as the superior reticular formation extend rostrally up to this level. A final deci-

12

sion on tegmental boundaries may result from homeotic gene expression data and grafting experiments. Recently, Marin and Puelles (1994) have demonstrated an embryonic polarizing gradient from the (rhombencephalic) isthmus region acting on the differentiation of mesencephalic structures in the avian brain. Their experiments indicate a mesencephalic-rhombencephalic boundary lying somewhat more rostral, i.e. between the oculomotor and trochlear motor nuclei and rostral to the interpeduncular nucleus.

Rhombencephalon (metencephalon and myelencephalon)

The terminology of Nieuwenhuys and Pouwels (1983) is applied except where noted.

The rhombencephalon (hindbrain) is often divided into a rostral metencephalon and a caudal myelencephalon. With the exception of the cerebellum, the ventral (medullary) remainder of the metencephalon can be separated only arbitrarily from the more caudal myelencephalic portion of the medulla oblongata. Thus, we treat cerebellum and medulla oblongata as entities here. Medulla oblongata and tegmentum are collectively referred to as brain stem.

The terms metencephalon and myelencephalon are only meaningful in mammals and birds. In those derived vertebrates, the metencephalon appears to be clearly separable from the myelencephalon as it exhibits a large dorsal cerebellum and ventral pons, which consists of relay neurons for cortical fibers to the cerebellum.

Cerebellum

As in all teleosts, the cerebellum (Ce) of the zebrafish has three parts: the vestibulolateralis lobe (including the medial caudal lobe, LCa, and the paired lateral eminentiae granulares, EG), the corpus cerebelli (CCe), and the valvula cerebelli, which has medial and lateral subdivisions (Vam/Val). The cerebellar commissure (Ccer) is located

within the ventral boundary zone between valvula and corpus cerebelli.

Although the valvula extends into the tectal ventricle, its histology (presence of a granular and a molecular layer, plus aggregations of large Purkinje- and eurydendroid cells) and its caudal attachment to the rostral medulla oblongata leave no doubt about it being part of the cerebellum. While both the vestibulolateralis lobe and the corpus cerebelli have homologues in other vertebrates, the valvula cerebelli is uniquely present in ray-finned fishes (Nieuwenhuys, 1967; Wullimann and Northcutt, 1988; 1989).

Medulla oblongata
Primary sensory and motor nuclei

The medulla oblongata (MO) contains the sensory and motor nuclei of the trigeminal (nervus trigeminus, V), abducens (nervus abducens, VI), facial (nervus facialis, VII), octaval (nervus octavus, VIII), glossopharyngeal (nervus glossopharyngeus, IX) and vagal (nervus vagus, X) nerves. The anterior and posterior lateral line nerves (ALLN/ PLLN) are separate from the other cranial nerves. In terms of the number of ganglia and peripheral innervation of neuromasts, both the anterior and posterior lateral line nerve roots include more than one nerve (Northcutt, 1989). However, these nerves are included here in the anterior and posterior lateral line nerve roots shown for the zebrafish (Fig. 4).

There are two separate trigeminal motor nuclei, one located dorsally to the lateral longitudinal fascicle (NVmd), and one situated at the ventrolateral edge of this fascicle (NVmv). Both motor nuclei extend more caudally where the lateral londitudinal fascicle runs more medially. Four trigeminal sensory nuclei (Puzdrowski, 1988) are described here. The most rostral, the isthmic primary sensory trigeminal nucleus (NVs), lies immediately caudal to the secondary gustatory nucleus. More caudally, at the mediodorsal edge of the descending trigeminal root (DV), lies the less clearly delineated (sensory) nucleus of the descending trigeminal root (NDV). It can best be recognized at the level of the caudal octavolateralis region. A third trigeminal sensory nucleus, the medial funicular nucleus (MFN), emerges at the very end of the medulla oblongata. The large pyriform neurons of the mesencephalic nucleus of the trigeminal nerve (MNV; located at the tectal ventricle near the synencephalon and optic tectum) have sensory fibers running peripherally in the trigeminal nerve (see chapter 6: Motor nuclei of cranial nerves).

The abducens nerve has two separate populations of motor neurons. The rostral motor nucleus and its root (VIr) are located at the level of the superior reticular formation, and the caudal motor nucleus (NVIc) and its root (VIc) are at the level of the intermediate reticular formation. Once outside the brain stem, the roots fuse and course rostrally.

The sensory root of the facial nerve (VIIs) enters the brainstem together with the anterior lateral line nerves. The sensory facial root first courses towards the midline of the brain stem, where it turns caudally and finally terminates in the facial lobe (LVII). Ventral to the facial sensory root and dorsal to the ventral rhombencephalic commissure and intermediate reticular formation lies the facial motor nucleus (NVIIm).

The octaval nerve (VIII) enters the zebrafish brain in an extended rostro-caudal region. Five primary sensory nuclei receive its projections: the anterior, magnocellular, descending, and posterior octaval nuclei, plus the tangential nucleus. It is difficult to subdivide the octaval area in teleosts. Subdivisions of this area in the zebrafish were carried out in close comparison with a detailed connectional study in the goldfish (McCormick and Hernandez, 1996).

The large neurons of the tangential nucleus (T) are located at the periphery of the brain stem, immediately ventral to the entrance of the anterior portion of the octaval nerve. These darkly staining neurons have a very distinctive large nucleus and nucleolus, and are distributed over a distance of only about 100µm. The slightly smaller neurons located more caudally are part of the descending octaval nucleus (DON), which is by far the

largest octaval system. It not only extends ventrally and dorsally around the octaval root but also reaches far medially. Its most caudal part at the level of the vagal root is referred to as the posterior octaval nucleus (PON). The magnocellular octaval nucleus (MaON) is located between the tangential nucleus and the transversely coursing portion of the sensory facial root. Rostral to the magnocellular octaval nucleus, the anterior octaval nucleus (AON) arches dorsally around the descending trigeminal root. Dorsal to the inner arcuate fibers, another octaval-related nucleus (although not a primary sensory nucleus) is present, referred to as the secondary octaval population (SO).

The root of the glossopharyngeal nerve (IX) which is peripherally related to the first gill arch, enters the brain ventral to the secondary gustatory tract. While its sensory nucleus forms a cell aggregation (LIX) located intermediate between facial and vagal lobes, its motor neurons do not form a separate nucleus. Rather, they are contained in the visceromotor column together with the motor neurons of the vagal nerve (NXm). The vagal nerve innervates the remaining gill arches and is, therefore, of much larger diameter. It carries gustatory information which is processed in the vagal lobe (LX), a large, paired brain stem structure flanking the unpaired facial lobe laterally. While some degree of histological segregation is apparent within the vagal lobe, a clear lamination, as seen in goldfish or carp, is absent. Also, from histology alone it cannot be decided whether motor neurons are included in the vagal lobe or not. The viscerosensory commissural nucleus of Cajal (NC) lies caudal to the vagal lobe.

There are two sensory nuclei related to the lateral line nerves in the zebrafish: the medial and caudal octavolateralis nuclei. The extensive medial octavolateralis nucleus (MON) is located in the most dorsal brain stem above the octaval nuclei; it is covered by a molecular layer called the cerebellar crest (CC). The latter is conventionally considered part of the medulla oblongata and not of the cerebellum, although the granular eminence cells extend parallel fibers into the cerebellar crest. The caudal octavolateralis nucleus (CON) is smaller and lies lateral to the facial and vagal lobes.

Reticular formation
The rhombencephalic reticular formation can be divided into midline, medial, and lateral columns (Nieuwenhuys and Pouwels, 1983). Immediately caudal to the interpeduncular nucleus, the midline column includes the superior raphe nucleus (SR). Its large neuronal perikarya are surrounded by a distinct neuropil. This is not the case for the more irregularly spaced neurons of the inferior raphe nucleus (IR), which lie in the ventral midline of the brain stem at the level of the facial and vagal lobes. The medial column of the reticular formation includes the superior, intermediate, and inferior nuclei of the reticular formation, called here for convenience superior, intermediate, and inferior reticular formation (SRF/IMRF/IRF). The superior reticular formation extends rostrally into the mesencephalon. The lateral column of the reticular formation includes the cerebellar-projecting lateral reticular nucleus (LRN).

Additional medullary nuclei
The griseum centrale (GC) is a longitudinally oriented nucleus situated along the ventral lining of the rhombencephalic ventricle. It extends partially into the mesencephalon.

The locus coeruleus (LC) consists of a few conspicuously shaped, large neurons dorsal to the superior reticular formation. Their widespread noradrenergic projections have recently been documented in the zebrafish (Ma, 1994a; 1994b).

The nucleus lateralis valvulae (NLV) is a large collection of granular cells at the ventral border of the cerebellum and brain stem. The dorsal tegmental nucleus (DTN) is sometimes considered part of the NLV. However, its cells are more densely packed than are those of the NLV and extend rostrally into the mesencephalon.

Two higher-order sensory nuclei are clearly delineable in the rostral brain stem. The visually related nucleus isthmi (NI) and the secondary gustatory nucleus (SGN). The large Mauthner cell (MAC) lies in the rostral octavolateralis region, between the anterior octaval nucleus and the ventral rhombencephalic commissure. It has two large dendrites, a lateral one extending towards the anterior/magnocellular octaval nuclei and a ventral one extending into the intermediate reticular formation. The Mauthner axon (MA) crosses the midline within the dorsal part of the medial longitudinal fascicle (MLF), and this heavily myelinated axon then descends in the MLF into the spinal cord.

The inferior olive (IO) is a large nucleus at the ventral periphery of the caudal brain stem. It is the source of climbing fibers reaching the cerebellum in teleosts (Finger, 1983; Wullimann and Northcutt, 1988; 1989).

Medulla spinalis

The terminology of Nieuwenhuys and Pouwels (1983) is applied except where noted. Only the rostral spinal cord at the level of the entrance of the second dorsal root (DR) is characterized here. The second dorsal root is treated here, because the first dorsal root is minute. (Note, however, that the corresponding first ventral root is huge and likely innervates hypaxial (somatic) musculature in the lower jaw.) First dorsal and ventral spinal roots are located approximately 100–150 µm caudal to the commissura infima of Haller. The second dorsal root, shown in this atlas, lies about 500–800 µm caudal to that commissure.

At the level of the second dorsal root, dorsal and ventral horns (DH/VH) of the grey matter are clearly visible surrounding the central canal (C). The longitudinally running tracts have been rearranged dramatically compared to their position and composition in the brain stem. They lie in the peripherally located white matter. The white matter of the spinal cord can be subdivided here into dorsal, lateral (which consists of a dorsal and a ventral part), and ventral funiculi (Fd/Fld/Flv/Fv), as in mammalian neuroanatomy (Nieuwenhuys et al., 1988).

Brain stem/spinal tracts and commissures

The terminology is according to Nieuwenhuys and Pouwels (1983) unless otherwise indicated.

Approximately at midtectal levels, the descending fibers of the optic tectum form the massive tecto-bulbar tract (TTB) running ventromedially along the surface of the torus semicircularis and the tegmentum. Many fibers in the tecto-bulbar tract cross the midline in the ansulate commissure (Cans), which is located immediately rostral to the interpeduncular nucleus. Lateral to this nucleus, the crossed tecto-bulbar tract (TTBc) then continues its caudal course into the ventromedial brain stem. A portion of the uncrossed tecto-bulbar tract (= tractus tectobulbaris rectus, TTBr) detaches very rostrally from the main tecto-bulbar fiber masses and runs caudally, separate from the more laterally located main portion of the uncrossed tecto-bulbar tract. At the level of the inferior reticular formation, both TTBc and TTBr are no longer visible.

Because of its large size, the medial longitudinal fascicle (MLF) is often considered to be the major descending fiber system in the brain of anamniotes. It begins within the nucleus of the MLF and runs – immediately ventral to the rhombencephalic ventricle (RV) – towards the caudal end of the medulla oblongata. Whereas the dorsal part of the MLF (including the Mauthner axon) continues to course into the spinal cord, the ventral portion of the MLF associates with other fiber systems in the funiculus ventralis prior to reaching the spinal cord. Throughout most of its rhombencephalic extent, the MLF is intersected by the ventral rhombencephalic commissure (Cven). A smaller commissure, the octavolateralis-related inner arcuate fibers (IAF), also crosses via the MLF.

The ascending lateral longitudinal fascicle (LLF) is the homologue of the mammalian lateral lemniscus. The LLF contains fibers of the auditory and mechanosensory sys-tems which originate in the primary sensory brainstem nuclei and terminate in the torus semicircularis. Caudally, the LLF lies dorsal to the superior reticular formation, adjacent to the ventral rhombencephalic commissure. It becomes displaced more laterally as it approaches the torus semicircularis.

The anterior mesencephalo-cerebellar tract (TMCa) carries mostly cerebellar afferents from the pretectum (Wullimann and Northcutt, 1988; 1989). The TMCa runs medial to the lateral longitudinal fascicle for some distance before turning dorsally to traverse the nucleus lateralis valvulae. Upon entering the cerebellum, part of the TMCa fuses with the posterior mesencephalo-cerebellar tract (TMCp), which contains the axons of the dorsal tegmental nucleus and nucleus lateralis valvulae. The TMCa and TMCp together form the anterior cerebellar tract (AC). A posterior cerebellar tract (PC) carrying cerebellar afferents from many brain stem nuclei (Wullimann and Northcutt, 1988; 1989) enters the cerebellar corpus at the level of the granular eminence. Many efferent cerebellar fibers decussate in the brachium conjunctivum and terminate, for example, in the nucleus ruber (Wullimann and Northcutt, 1988). In the zebrafish we could not identify the brachium conjunctivum neuroanatomically.

On entering the brain stem, the sensory root of the trigeminal nerve (Vs) bifurcates. Besides giving off fibers to the isthmic primary sensory trigeminal nucleus (NVs; Puzdrowski, 1988), the sensory trigeminal root turns caudally and becomes the descending trigeminal root (DV). Located ventral to the octavolateralis area, the DV descends towards the caudal tip of the medulla oblongata. Here, the DV is located lateral to the medial funicular nucleus, which receives a considerable trigeminal input. Those fibers which eventually project further caudally into the spinal cord become located within the dorsal part of the lateral funiculus.

Throughout most of its course, the descending trigeminal root is accompanied ventromedially by the ascending secondary gustatory tract (SGT), which runs from the primary gustatory centers to the secondary gustatory nucleus.

The bulbo-spinal tract (TBS) emerges medial to the caudal intermediate reticular formation and runs caudally, constantly growing in size, along the medial edge of the inferior reticular formation. Immediately prior to reaching the spinal cord, the bulbo-spinal tract is displaced laterally and associates with other tracts in the dorsal part of the funiculus lateralis.

The vestibulo-spinal tract (TVS) forms at the rostral level of the inferior reticular formation where it lies dorsal to the inferior olive. It associates more caudally with other tracts in the ventral part of the funiculus lateralis. Both the TVS and the TBS carry descending spinal projections. Two additional commissures are present in the medulla oblongata. The commissure of the secondary gustatory nuclei (Cgus; Herrick, 1905) runs between these large, paired sensory nuclei. The commissura infima of Haller (Cinf) is located dorsal to the commissural nucleus of Cajal.

5 The brain of the zebrafish *Danio rerio*: a neuroanatomical atlas

External view

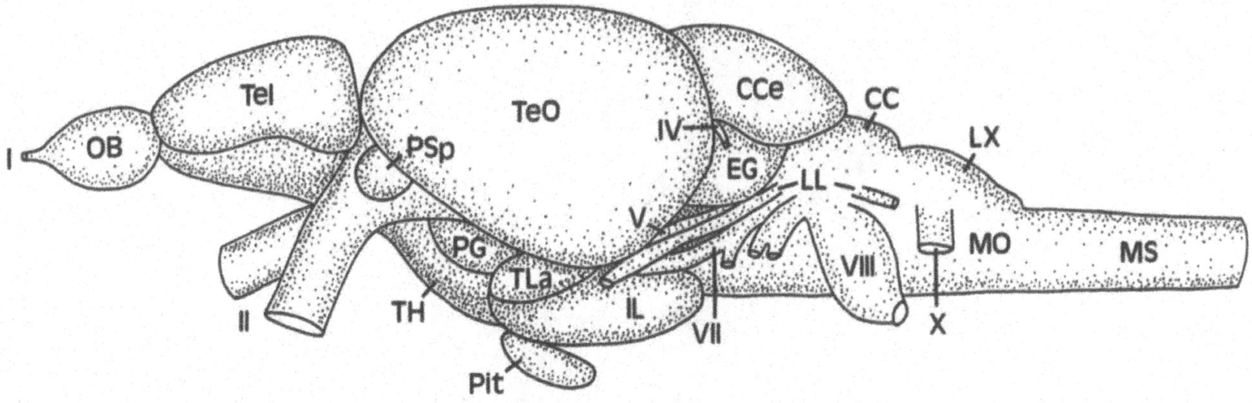

Figure 4.
Lateral (a) and dorsal (b) views of the adult zebrafish brain. The telencephalon comprises a dorsal and a ventral telencephalic area as well as an olfactory bulb, which is entered rostrally by the olfactory nerve. The diencephalon is located rostroventral to the midbrain and is largely covered by the optic tectum. Externally visible portions of the diencephalon include the optic nerve, the preglomerular area, the torus lateralis, the habenula, and the hypothalamus. The hypothalamus can be separated into a rostromedial tuberal hypothalamus and paired lateral lobes. Saccus dorsalis and epiphysis were removed during preparation and are not shown. The brainstem, which harbors most cranial nerve roots except for the optic and olfactory nerves, is covered rostrally by the cerebellum (i.e. the corpus cerebelli and the granular eminence), and includes more caudally the crista cerebellaris, the prominent paired vagal lobes, and the unpaired facial lobe. Finally, the medulla oblongata grades into the spinal cord (medulla spinalis).

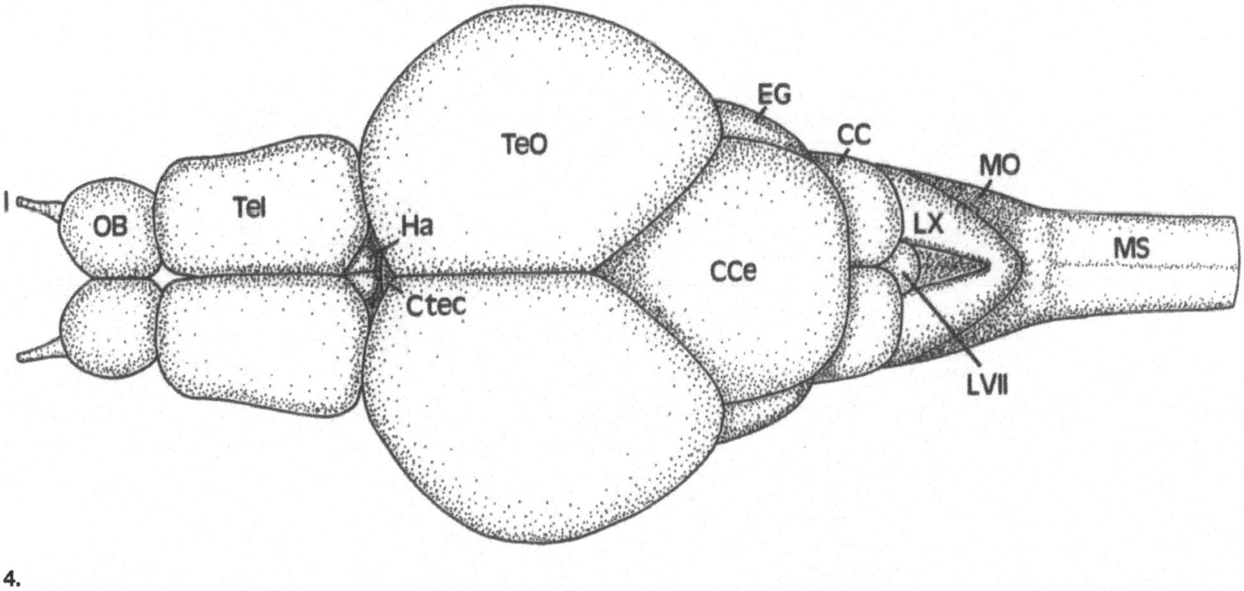

CC	crista cerebellaris	PSp	parvocellular superficial
CCe	corpus cerebelli		pretectal nucleus
Ctec	commissura tecti	Tel	telencephalon
EG	eminentia granularis	TeO	tectum opticum
Ha	habenula	TH	tuberal hypothalamus
IL	inferior lobe of	TLa	torus lateralis
	hypothalamus		
LL	lateral line nerves	I	olfactory nerve
LVII	facial lobe	II	optic nerve
LX	vagal lobe	IV	trochlear nerve
MO	medulla oblongata	V	trigeminal nerve
MS	medulla spinalis	VII	facial nerve
OB	olfactory bulb	VIII	octaval nerve
PG	preglomerular area	X	vagal nerve
Pit	pituitary		

Cross Sections

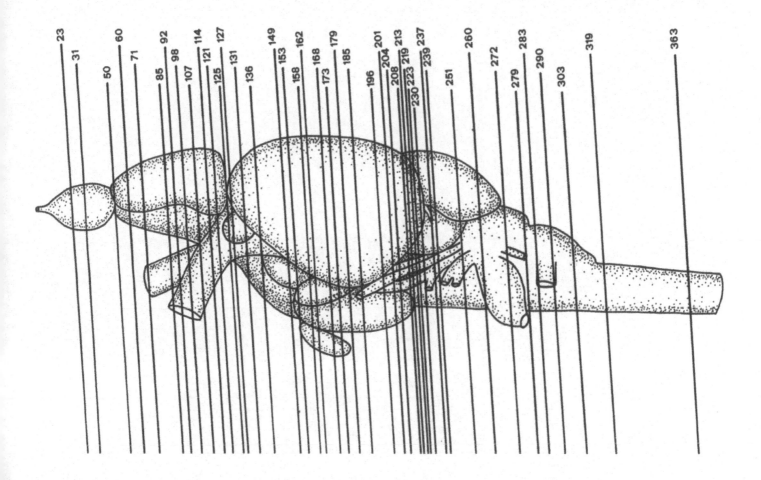

Figure 5.

Lateral view of the adult zebrafish brain indicating the position of levels illustrated in the following series of cross sections. This series gives a detailed overview of the neuroanatomy of the adult zebrafish brain. Individual sections were not taken at equal intervals but were chosen according to the appearance of novel anatomical structures, beginning from the rostralmost extension of the olfactory bulb and proceeding through the entire brain up to the rostral spinal cord. The section plane of this series is bent slightly rostroventrally rather than corresponding to the ideal vertical axis.

21

ECL external cellular layer of olfactory
     bulb including mitral cells
GL   glomerular layer of olfactory bulb
POF primary olfactory fiber layer

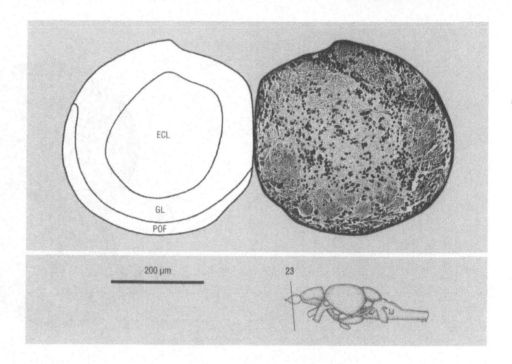

ECL external cellular layer of olfactory
     bulb including mitral cells
GL   glomerular layer of olfactory bulb
ICL internal cellular layer of olfactory
     bulb

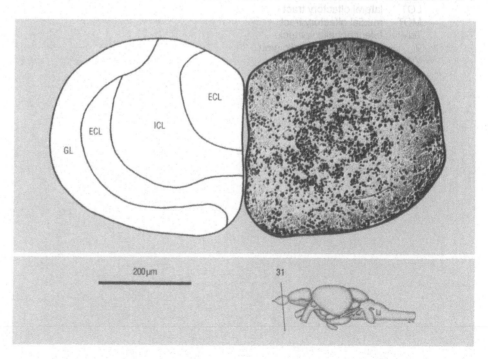

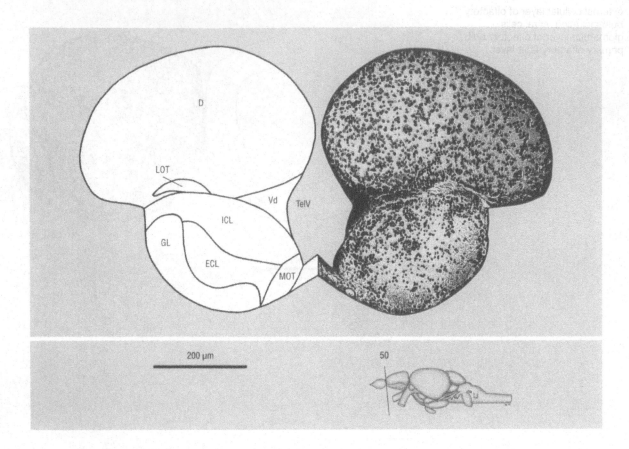

200 µm

50

D   dorsal telencephalic area
ECL   external cellular layer of olfactory bulb including mitral cells
GL   glomerular layer of olfactory bulb
ICL   internal cellular layer of olfactory bulb
LOT   lateral olfactory tract
MOT   medial olfactory tract
TelV   telencephalic ventricle
V   ventral telencephalic area
Vd   dorsal nucleus of V

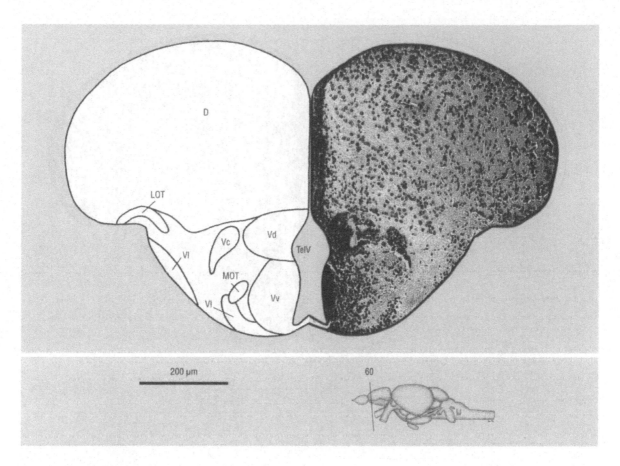

200 μm

60

D	dorsal telencephalic area
LOT	lateral olfactory tract
MOT	medial olfactory tract
TelV	telencephalic ventricle
V	ventral telencephalic area
Vc	central nucleus of V
Vd	dorsal nucleus of V
Vl	lateral nucleus of V
Vv	ventral nucleus of V

Cross Section 71

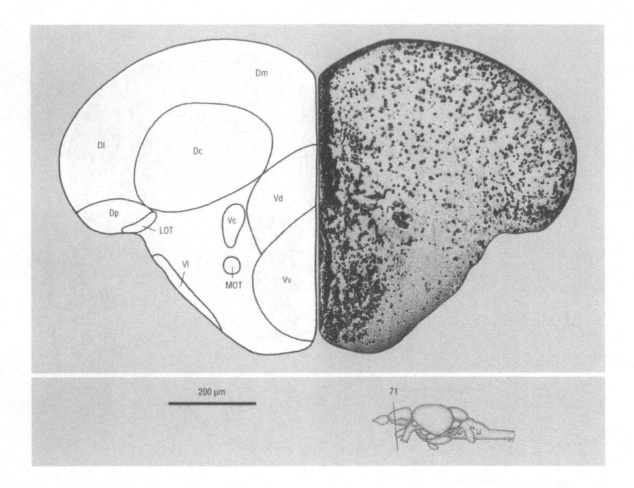

D	dorsal telencephalic area
Dc	central zone of D
Dl	lateral zone of D
Dm	medial zone of D
Dp	posterior zone of D
LOT	lateral olfactory tract
MOT	medial olfactory tract
V	ventral telencephalic area
Vc	central nucleus of V
Vd	dorsal nucleus of V
Vl	lateral nucleus of V
Vv	ventral nucleus of V

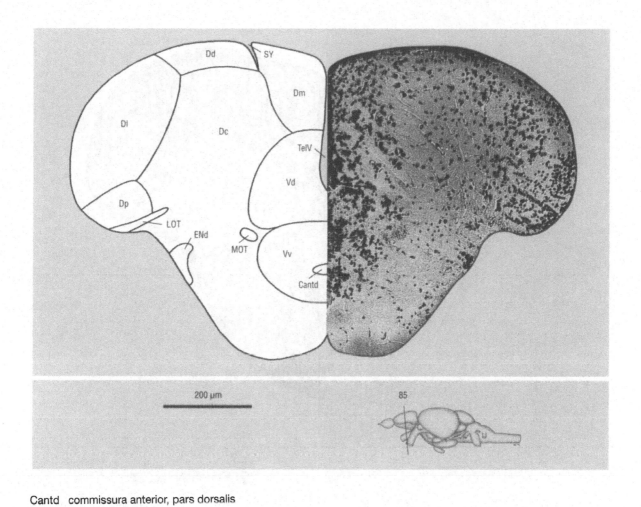

200 μm

85

Cantd	commissura anterior, pars dorsalis
D	dorsal telencephalic area
Dc	central zone of D
Dd	dorsal zone of D
Dl	lateral zone of D
Dm	medial zone of D
Dp	posterior zone of D
ENd	entopeduncular nucleus, dorsal part
LOT	lateral olfactory tract
MOT	medial olfactory tract
SY	sulcus ypsiloniformis
TelV	telencephalic ventricle
V	ventral telencephalic area
Vd	dorsal nucleus of V
Vv	ventral nucleus of V

Cross Section 92

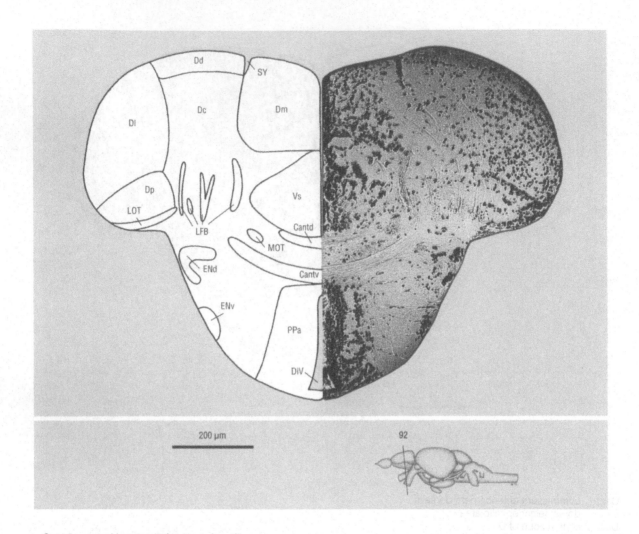

200 µm

92

Cantd	commissura anterior, pars dorsalis
Cantv	commissura anterior, pars ventralis
D	dorsal telencephalic area
Dc	central zone of D
Dd	dorsal zone of D
DiV	diencephalic ventricle
Dl	lateral zone of D
Dm	medial zone of D
Dp	posterior zone of D
ENd	entopeduncular nucleus, dorsal part
ENv	entopeduncular nucleus, ventral part
LFB	lateral forebrain bundle
LOT	lateral olfactory tract
MOT	medial olfactory tract
PPa	parvocellular preoptic nucleus, anterior part
SY	sulcus ypsiloniformis
V	ventral telencephalic area
Vs	supracommissural nucleus of V

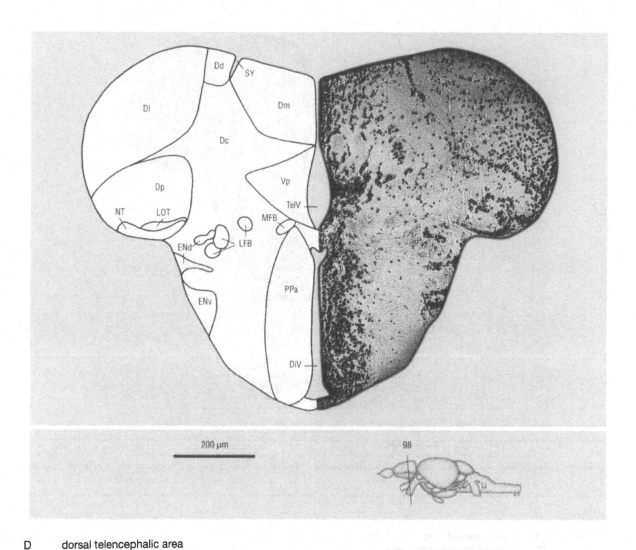

200 μm

98

D	dorsal telencephalic area
Dc	central zone of D
Dd	dorsal zone of D
DiV	diencephalic ventricle
Dl	lateral zone of D
Dm	medial zone of D
Dp	posterior zone of D
ENd	entopeduncular nucleus, dorsal part
ENv	entopeduncular nucleus, ventral part
LFB	lateral forebrain bundle
LOT	lateral olfactory tract
MFB	medial forebrain bundle
NT	nucleus taeniae
PPa	parvocellular preoptic nucleus, anterior part
SY	sulcus ypsiloniformis
TelV	telencephalic ventricle
V	ventral telencephalic area
Vp	postcommissural nucleus of V

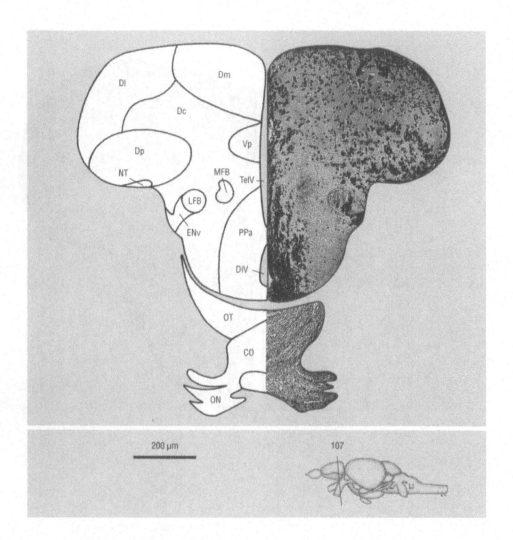

200 µm

107

CO	chiasma opticum
D	dorsal telencephalic area
Dc	central zone of D
DiV	diencephalic ventricle
Dl	lateral zone of D
Dm	medial zone of D
Dp	posterior zone of D
ENv	entopeduncular nucleus, ventral part
LFB	lateral forebrain bundle
MFB	medial forebrain bundle
NT	nucleus taeniae
ON	optic nerve
OT	optic tract
PPa	parvocellular preoptic nucleus, anterior part
TelV	telencephalic ventricle
V	ventral telencephalic area
Vp	postcommissural nucleus of V

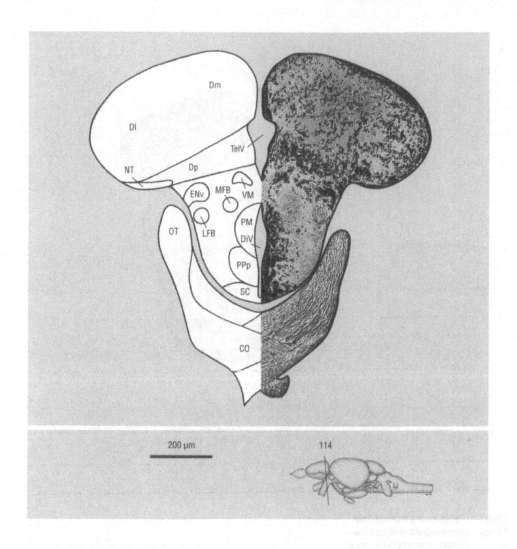

200 µm

114

CO	chiasma opticum
D	dorsal telencephalic area
DiV	diencephalic ventricle
Dl	lateral zone of D
Dm	medial zone of D
Dp	posterior zone of D
ENv	entopeduncular nucleus, ventral part
LFB	lateral forebrain bundle
MFB	medial forebrain bundle
NT	nucleus taeniae
OT	optic tract
PM	magnocellular preoptic nucleus
PPp	parvocellular preoptic nucleus, posterior part
SC	suprachiasmatic nucleus
TelV	telencephalic ventricle
VM	ventromedial thalamic nucleus

Cross Section 121

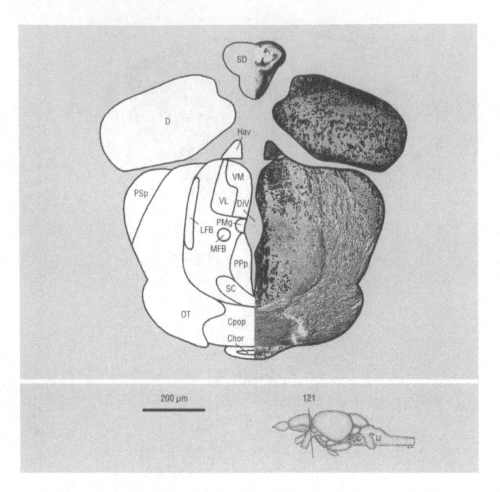

Chor	commissura horizontalis
Cpop	commissura postoptica
D	dorsal telencephalic area
DiV	diencephalic ventricle
Hav	ventral habenular nucleus
LFB	lateral forebrain bundle
MFB	medial forebrain bundle
OT	optic tract
PMg	gigantocellular part of magnocellular preoptic nucleus
PPp	parvocellular preoptic nucleus, posterior part
PSp	parvocellular superficial pretectal nucleus
SC	suprachiasmatic nucleus
SD	saccus dorsalis
VL	ventrolateral thalamic nucleus
VM	ventromedial thalamic nucleus

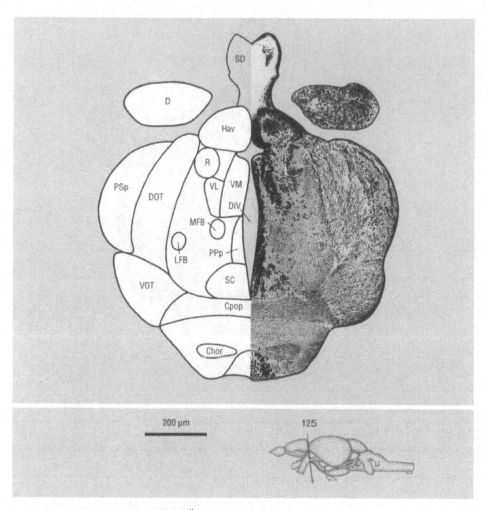

200 µm

125

Chor	commissura horizontalis
Cpop	commissura postoptica
D	dorsal telencephalic area
DiV	diencephalic ventricle
DOT	dorsomedial optic tract
Hav	ventral habenular nucleus
LFB	lateral forebrain bundle
MFB	medial forebrain bundle
PPp	parvocellular preoptic nucleus, posterior part
PSp	parvocellular superficial pretectal nucleus
R	rostrolateral nucleus (of Butler & Saidel 91)
SC	suprachiasmatic nucleus
SD	saccus dorsalis
VL	ventrolateral thalamic nucleus
VM	ventromedial thalamic nucleus
VOT	ventrolateral optic tract

Cross Section 127

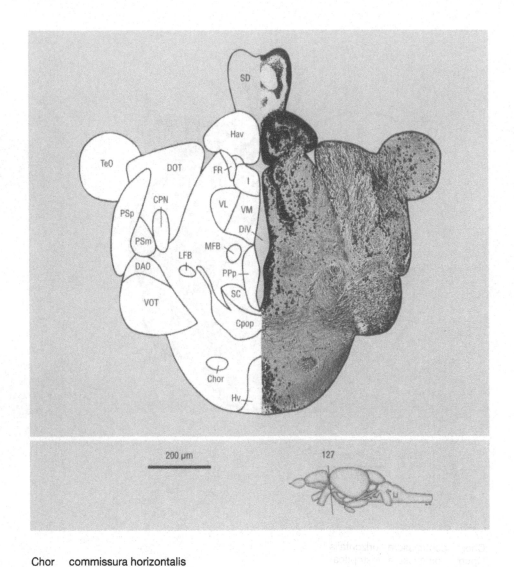

200 µm

127

Chor	commissura horizontalis
CPN	central pretectal nucleus
Cpop	commissura postoptica
DAO	dorsal accessory optic nucleus
DiV	diencephalic ventricle
DOT	dorsomedial optic tract
FR	fasciculus retroflexus
Hav	ventral habenular nucleus
Hv	ventral zone of periventricular hypothalamus
I	intermediate thalamic nucleus
LFB	lateral forebrain bundle
MFB	medial forebrain bundle
PPp	parvocellular preoptic nucleus, posterior part
PSm	magnocellular superficial pretectal nucleus
PSp	parvocellular superficial pretectal nucleus
SC	suprachiasmatic nucleus
SD	saccus dorsalis
TeO	tectum opticum
VL	ventrolateral thalamic nucleus
VM	ventromedial thalamic nucleus
VOT	ventrolateral optic tract

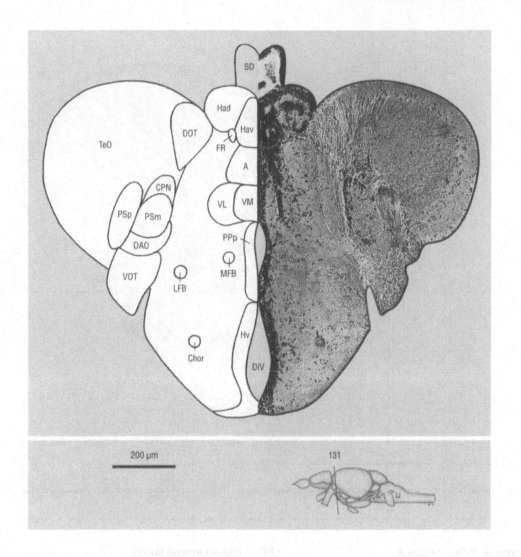

200 µm

131

A	anterior thalamic nucleus
Chor	commissura horizontalis
CPN	central pretectal nucleus
DAO	dorsal accessory optic nucleus
DiV	diencephalic ventricle
DOT	dorsomedial optic tract
FR	fasciculus retroflexus
Had	dorsal habenular nucleus
Hav	ventral habenular nucleus
Hv	ventral zone of periventricular hypothalamus
LFB	lateral forebrain bundle
MFB	medial forebrain bundle
PPp	parvocellular preoptic nucleus, posterior part
PSm	magnocellular superficial pretectal nucleus
PSp	parvocellular superficial pretectal nucleus
SD	saccus dorsalis
TeO	tectum opticum
VL	ventrolateral thalamic nucleus
VM	ventromedial thalamic nucleus
VOT	ventrolateral optic tract

Cross Section 136

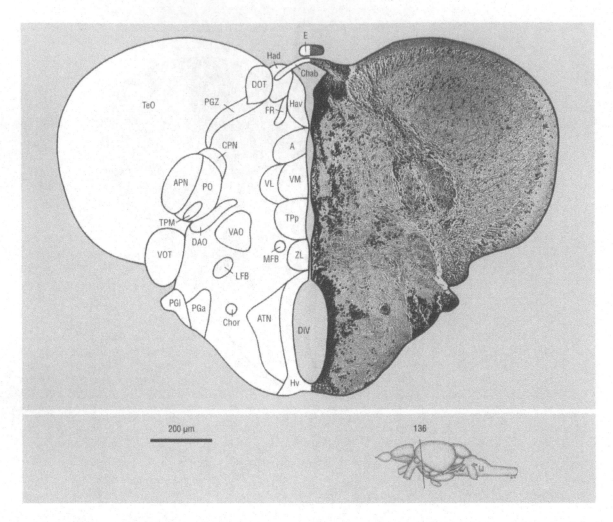

200 µm

136

A	anterior thalamic nucleus	LFB	lateral forebrain bundle
APN	accessory pretectal nucleus (of Wullimann & Meyer 90)	MFB	medial forebrain bundle
		PGa	anterior preglomerular nucleus
ATN	anterior tuberal nucleus	PGl	lateral preglomerular nucleus
Chab	commissura habenularum	PGZ	periventricular gray zone of optic tectum
Chor	commissura horizontalis	PO	posterior pretectal nucleus (of Wullimann & Meyer 90)
CPN	central pretectal nucleus		
DAO	dorsal accessory optic nucleus	TeO	tectum opticum
DiV	diencephalic ventricle	TPM	tractus pretectomamillaris
DOT	dorsomedial optic tract	TPp	periventricular nucleus of posterior tuberculum
E	epiphysis	VAO	ventral accessory optic nucleus
FR	fasciculus retroflexus	VL	ventrolateral thalamic nucleus
Had	dorsal habenular nucleus	VM	ventromedial thalamic nucleus
Hav	ventral habenular nucleus	VOT	ventrolateral optic tract
Hv	ventral zone of periventricular hypothalamus	ZL	zona limitans

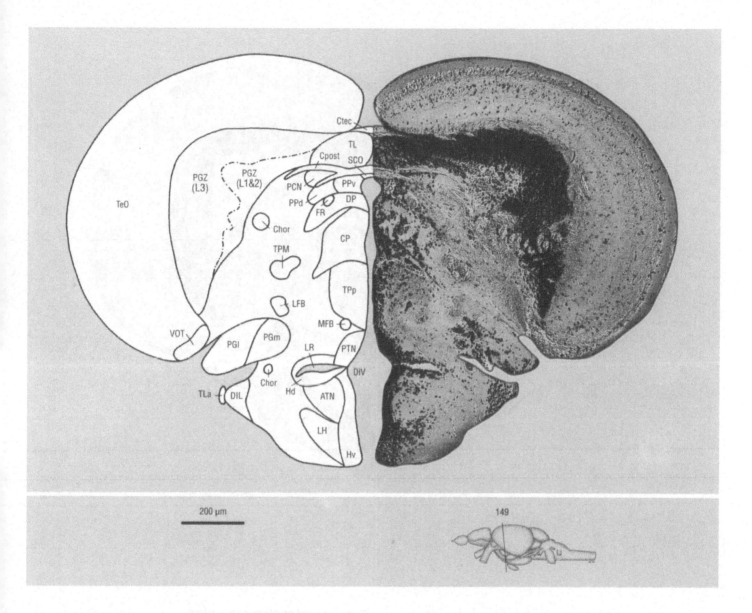

200 µm

149

ATN	anterior tuberal nucleus	PCN	paracommissural nucleus
Chor	commissura horizontalis	PGl	lateral preglomerular nucleus
CP	central posterior thalamic nucleus	PGm	medial preglomerular nucleus
Cpost	commissura posterior	PGZ	periventricular gray zone of optic tectum
Ctec	commissura tecti	PPd	periventricular pretectal nucleus, dorsal part
DIL	diffuse nucleus of the inferior lobe	PPv	periventricular pretectal nucleus, ventral part
DiV	diencephalic ventricle	PTN	posterior tuberal nucleus
DP	dorsal posterior thalamic nucleus	SCO	subcommissural organ
FR	fasciculus retroflexus	TeO	tectum opticum
Hd	dorsal zone of periventricular hypothalamus	TL	torus longitudinalis
Hv	ventral zone of periventricular hypothalamus	TLa	torus lateralis
LFB	lateral forebrain bundle	TPM	tractus pretectomamillaris
LH	lateral hypothalamic nucleus	TPp	periventricular nucleus of posterior tuberculum
LR	lateral recess of diencephalic ventricle	VOT	ventrolateral optic tract
MFB	medial forebrain bundle		

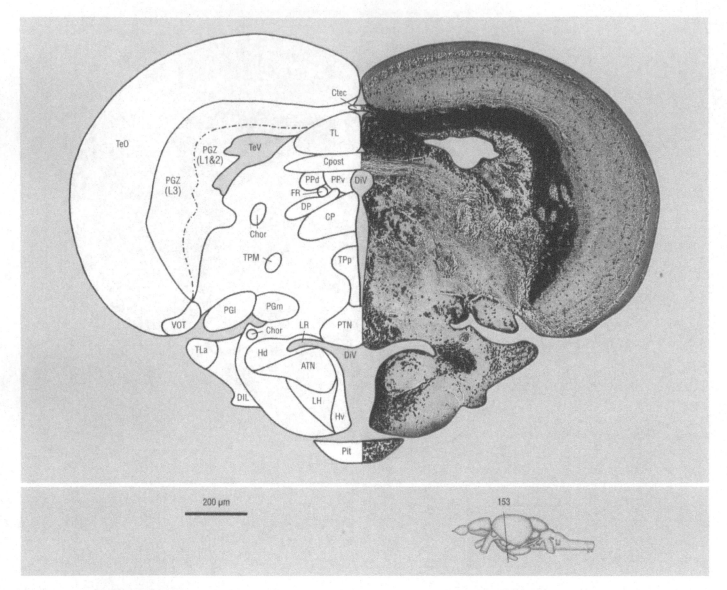

200 μm

153

ATN	anterior tuberal nucleus	PGm	medial preglomerular nucleus
Chor	commissura horizontalis	PGZ	periventricular gray zone of optic tectum
CP	central posterior thalamic nucleus	Pit	pituitary
Cpost	commissura posterior	PPd	periventricular pretectal nucleus, dorsal part
Ctec	commissura tecti	PPv	periventricular pretectal nucleus, ventral part
DIL	diffuse nucleus of the inferior lobe	PTN	posterior tuberal nucleus
DiV	diencephalic ventricle	TeO	tectum opticum
DP	dorsal posterior thalamic nucleus	TeV	tectal ventricle
FR	fasciculus retroflexus	TL	torus longitudinalis
Hd	dorsal zone of periventricular hypothalamus	TLa	torus lateralis
Hv	ventral zone of periventricular hypothalamus	TPM	tractus pretectomamillaris
LH	lateral hypothalamic nucleus	TPp	periventricular nucleus of posterior tuberculum
LR	lateral recess of diencephalic ventricle	VOT	ventrolateral optic tract
PGl	lateral preglomerular nucleus		

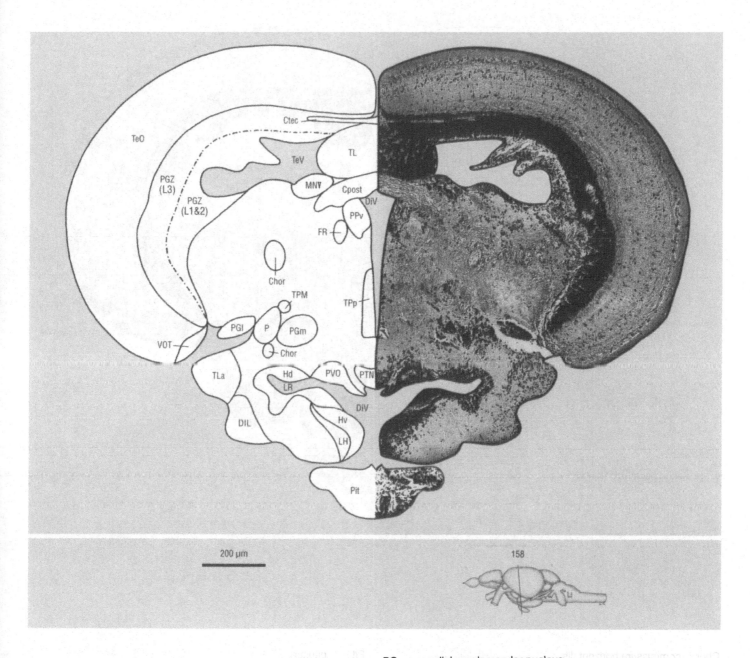

200 μm

158

Chor	commissura horizontalis	PGm	medial preglomerular nucleus
Cpost	commissura posterior	PGZ	periventricular gray zone of optic tectum
Ctec	commissura tecti	Pit	pituitary
DIL	diffuse nucleus of the inferior lobe	PPv	periventricular pretectal nucleus, ventral part
DiV	diencephalic ventricle	PTN	posterior tuberal nucleus
FR	fasciculus retroflexus	PVO	paraventricular organ
Hd	dorsal zone of periventricular hypothalamus	TeO	tectum opticum
Hv	ventral zone of periventricular hypothalamus	TeV	tectal ventricle
LH	lateral hypothalamic nucleus	TL	torus longitudinalis
LR	lateral recess of diencephalic ventricle	TLa	torus lateralis
MNV	mesencephalic nucleus of trigeminal nerve	TPM	tractus pretectomamillaris
P	posterior thalamic nucleus	TPp	periventricular nucleus of posterior tuberculum
PGl	lateral preglomerular nucleus	VOT	ventrolateral optic tract

Cross Section 162

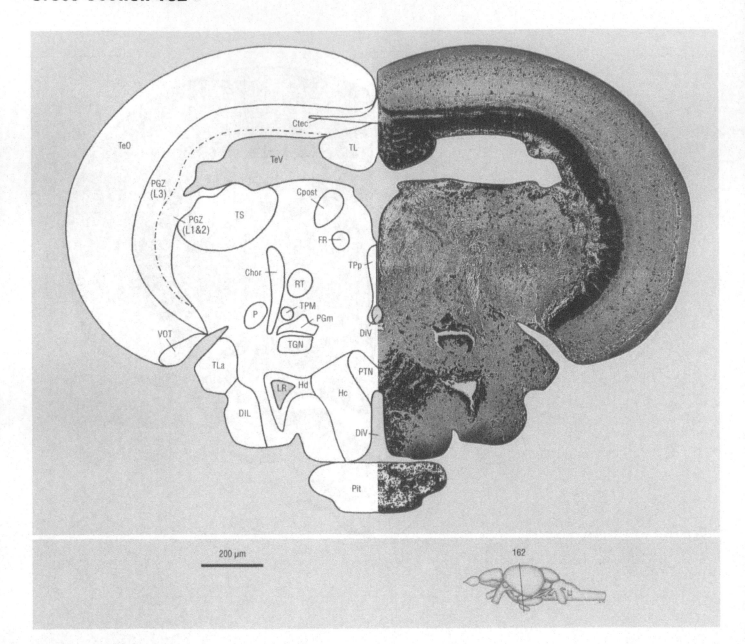

200 µm

162

Chor	commissura horizontalis	Pit	pituitary
Cpost	commissura posterior	PTN	posterior tuberal nucleus
Ctec	commissura tecti	RT	rostral tegmental nucleus (of Grover & Sharma 81)
DIL	diffuse nucleus of the inferior lobe	TeO	tectum opticum
DiV	diencephalic ventricle	TeV	tectal ventricle
FR	fasciculus retroflexus	TGN	tertiary gustatory nucleus (of Wullimann 88)
Hc	caudal zone of periventricular hypothalamus	TL	torus longitudinalis
Hd	dorsal zone of periventricular hypothalamus	TLa	torus lateralis
LR	lateral recess of diencephalic ventricle	TPM	tractus pretectomamillaris
P	posterior thalamic nucleus	TPp	periventricular nucleus of posterior tuberculum
PGm	medial preglomerular nucleus	TS	torus semicircularis
PGZ	periventricular gray zone of optic tectum	VOT	ventrolateral optic tract

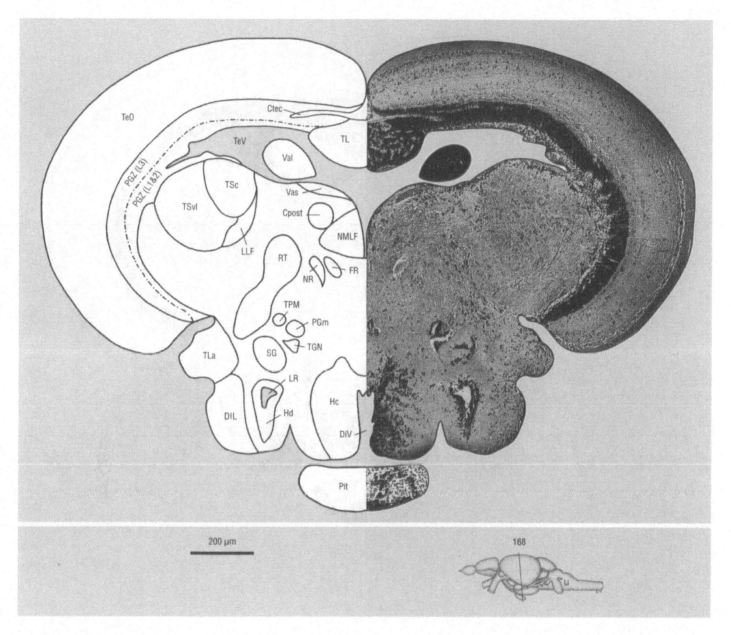

200 µm

168

Cpost	commissura posterior
Ctec	commissura tecti
DIL	diffuse nucleus of the inferior lobe
DiV	diencephalic ventricle
FR	fasciculus retroflexus
Hc	caudal zone of periventricular hypothalamus
Hd	dorsal zone of periventricular hypothalamus
LLF	lateral longitudinal fascicle
LR	lateral recess of diencephalic ventricle
NMLF	nucleus of MLF
NR	nucleus ruber
PGm	medial preglomerular nucleus
PGZ	periventricular gray zone of optic tectum

Pit	pituitary
RT	rostral tegmental nucleus (of Grover & Sharma 81)
SG	subglomerular nucleus
TeO	tectum opticum
TeV	tectal ventricle
TGN	tertiary gustatory nucleus (of Wullimann 88)
TL	torus longitudinalis
TLa	torus lateralis
TPM	tractus pretectomamillaris
TSc	central nucleus of torus semicircularis
TSvl	ventrolateral nucleus of torus semicircularis
Val	lateral division of valvula cerebelli
Vas	vascular lacuna of area postrema

Cross Section 173

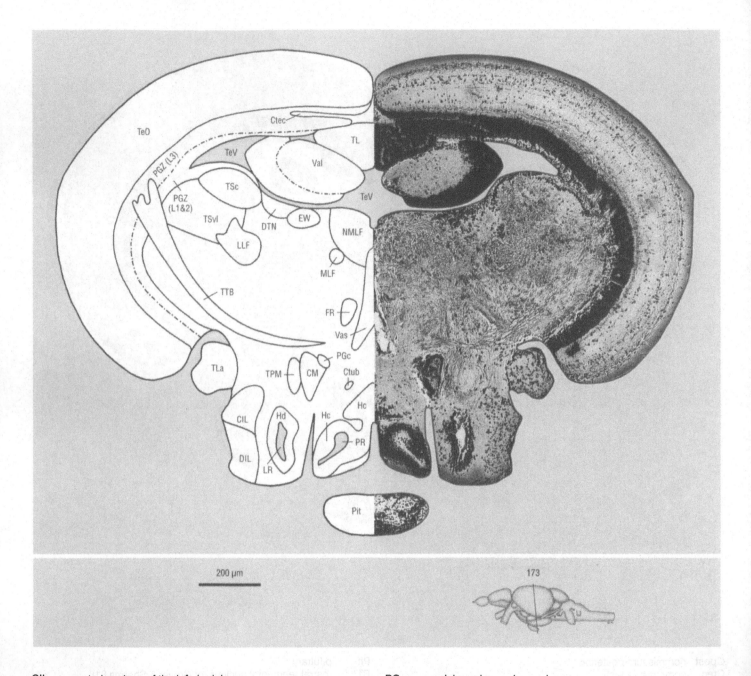

200 μm

173

CIL	central nucleus of the inferior lobe	PGc	caudal preglomerular nucleus
CM	corpus mamillare	PGZ	periventricular gray zone of optic tectum
Ctec	commissura tecti	Pit	pituitary
Ctub	commissure of the posterior tuberculum	PR	posterior recess of diencephalic ventricle
DIL	diffuse nucleus of the inferior lobe	TeO	tectum opticum
DTN	dorsal tegmental nucleus	TeV	tectal ventricle
EW	Edinger-Westphal nucleus	TL	torus longitudinalis
FR	fasciculus retroflexus	TLa	torus lateralis
Hc	caudal zone of periventricular hypothalamus	TPM	tractus pretectomamillaris
Hd	dorsal zone of periventricular hypothalamus	TSc	central nucleus of torus semicircularis
LLF	lateral longitudinal fascicle	TSvl	ventrolateral nucleus of torus semicircularis
LR	lateral recess of diencephalic ventricle	TTB	tractus tectobulbaris
MLF	medial longitudinal fascicle	Val	lateral division of valvula cerebelli
NMLF	nucleus of MLF	Vas	vascular lacuna of area postrema

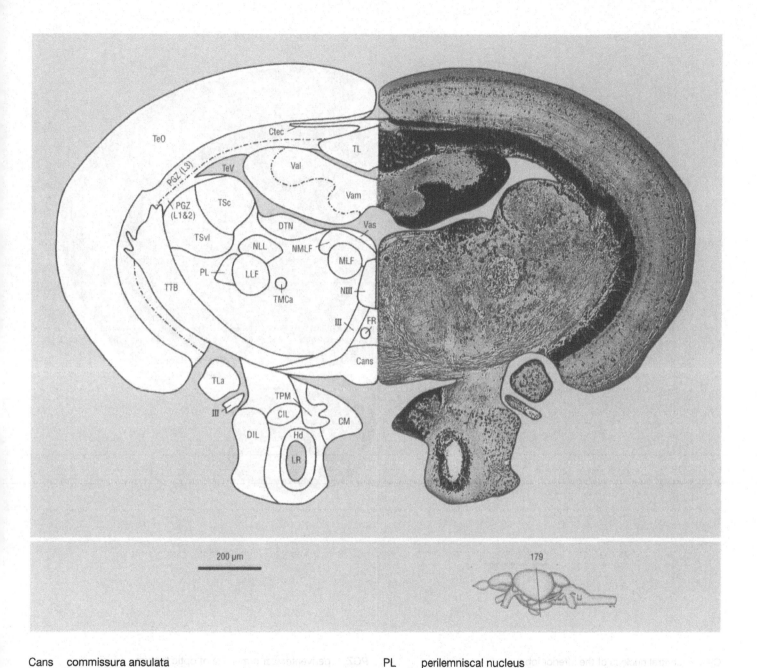

200 µm

179

Cans	commissura ansulata	PL	perilemniscal nucleus
CIL	central nucleus of the inferior lobe	TeO	tectum opticum
CM	corpus mamillare	TeV	tectal ventricle
Ctec	commissura tecti	TL	torus longitudinalis
DIL	diffuse nucleus of the inferior lobe	TLa	torus lateralis
DTN	dorsal tegmental nucleus	TMCa	tractus mesencephalocerebellaris anterior
FR	fasciculus retroflexus	TPM	tractus pretectomamillaris
Hd	dorsal zone of periventricular hypothalamus	TSc	central nucleus of torus semicircularis
LLF	lateral longitudinal fascicle	TSvl	ventrolateral nucleus of torus semicircularis
LR	lateral recess of diencephalic ventricle	TTB	tractus tectobulbaris
MLF	medial longitudinal fascicle	Val	lateral division of valvula cerebelli
NLL	nucleus of the lateral lemniscus (of Prasada Rao et al. 87)	Vam	medial division of valvula cerebelli
NMLF	nucleus of MLF	Vas	vascular lacuna of area postrema
NIII	oculomotor nucleus	III	oculomotor nerve
PGZ	periventricular gray zone of optic tectum		

Cross Section 185

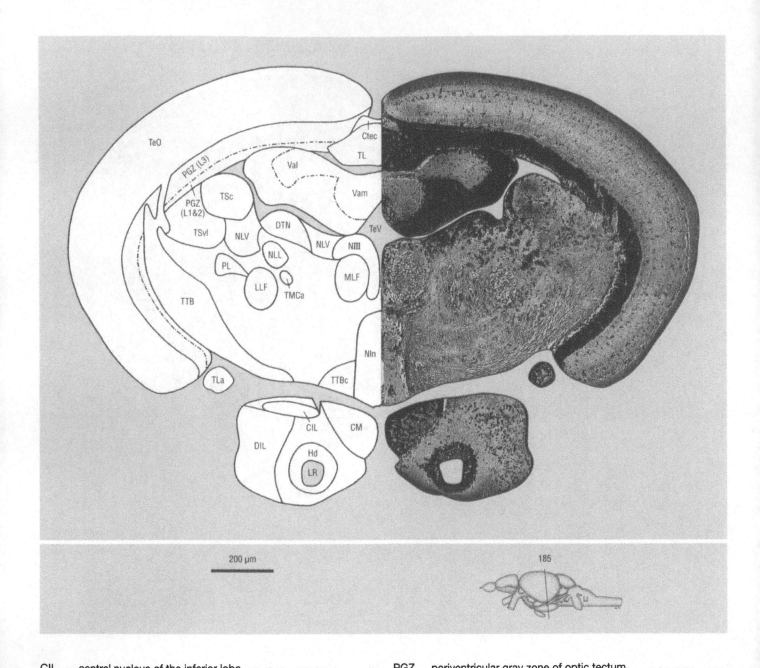

200 µm

185

CIL	central nucleus of the inferior lobe	PGZ	periventricular gray zone of optic tectum
CM	corpus mamillare	PL	perilemniscal nucleus
Ctec	commissura tecti	TeO	tectum opticum
DIL	diffuse nucleus of the inferior lobe	TeV	tectal ventricle
DTN	dorsal tegmental nucleus	TL	torus longitudinalis
Hd	dorsal zone of periventricular hypothalamus	TLa	torus lateralis
LLF	lateral longitudinal fascicle	TMCa	tractus mesencephalocerebellaris anterior
LR	lateral recess of diencephalic ventricle	TSc	central nucleus of torus semicircularis
MLF	medial longitudinal fascicle	TSvl	ventrolateral nucleus of torus semicircularis
NIn	nucleus interpeduncularis	TTB	tractus tectobulbaris
NLL	nucleus of the lateral lemniscus (of Prasada Rao et al. 87)	TTBc	tractus tectobulbaris cruciatus
NLV	nucleus lateralis valvulae	Val	lateral division of valvula cerebelli
NIII	oculomotor nucleus	Vam	medial division of valvula cerebelli

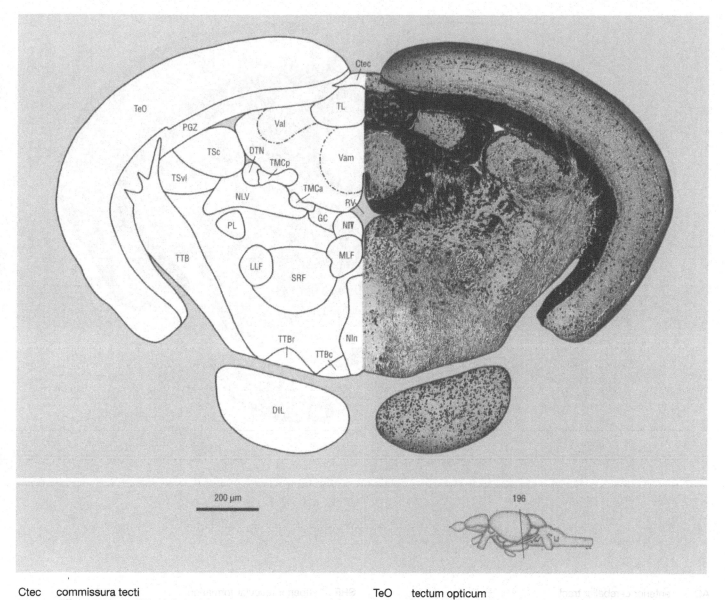

200 µm

196

Ctec	commissura tecti
DIL	diffuse nucleus of the inferior lobe
DTN	dorsal tegmental nucleus
GC	griseum centrale
LLF	lateral longitudinal fascicle
MLF	medial longitudinal fascicle
NIn	nucleus interpeduncularis
NLV	nucleus lateralis valvulae
NIV	trochlear nucleus
PGZ	periventricular gray zone of optic tectum
PL	perilemniscal nucleus
RV	rhombencephalic ventricle
SRF	superior reticular formation

TeO	tectum opticum
TeV	tectal ventricle
TL	torus longitudinalis
TMCa	tractus mesencephalocerebellaris anterior
TMCp	tractus mesencephalocerebellaris posterior
TSc	central nucleus of torus semicircularis
TSvl	ventrolateral nucleus of torus semicircularis
TTB	tractus tectobulbaris
TTBc	tractus tectobulbaris cruciatus
TTBr	tractus tectobulbaris rectus
Val	lateral division of valvula cerebelli
Vam	medial division of valvula cerebelli

Cross Section 201

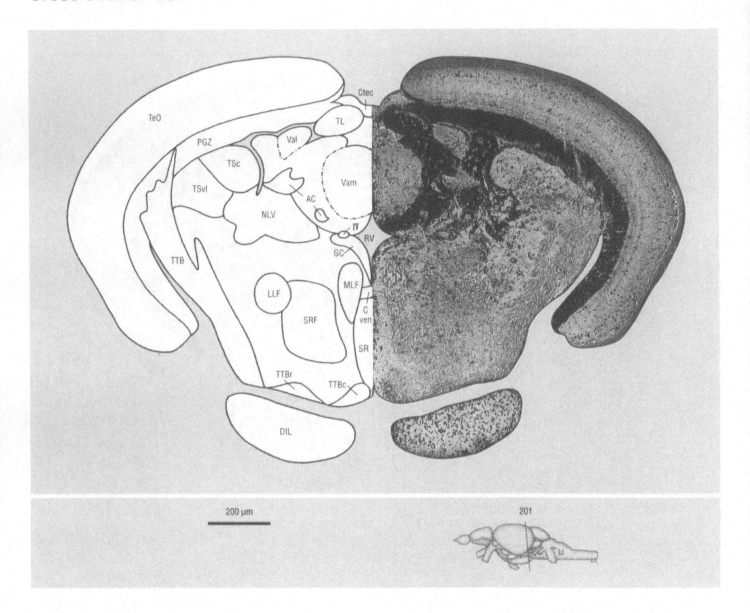

200 µm

201

AC	anterior cerebellar tract	SRF	superior reticular formation
Ctec	commissura tecti	TeO	tectum opticum
Cven	commissura ventralis rhombencephali	TL	torus longitudinalis
DIL	diffuse nucleus of the inferior lobe	TSc	central nucleus of torus semicircularis
GC	griseum centrale	TSvl	ventrolateral nucleus of torus semicircularis
LLF	lateral longitudinal fascicle	TTB	tractus tectobulbaris
MLF	medial longitudinal fascicle	TTBc	tractus tectobulbaris cruciatus
NLV	nucleus lateralis valvulae	TTBr	tractus tectobulbaris rectus
PGZ	periventricular gray zone of optic tectum	Val	lateral division of valvula cerebelli
RV	rhombencephalic ventricle	Vam	medial division of valvula cerebelli
SR	superior raphe	IV	trochlear nerve

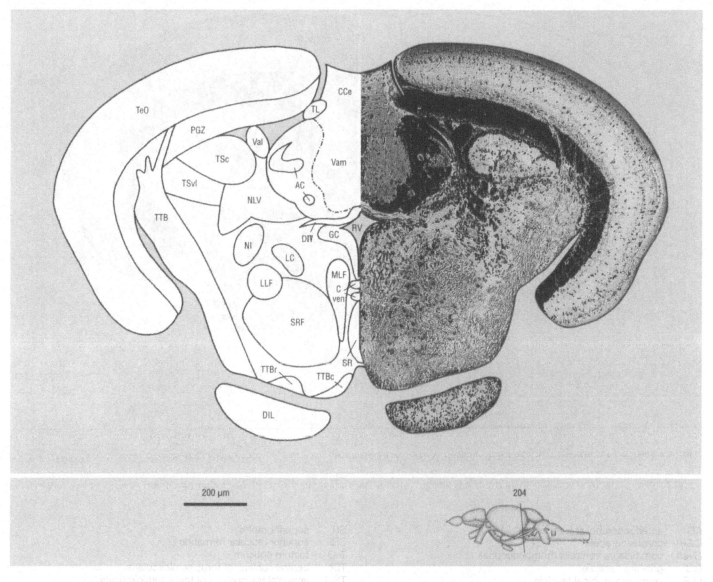

200 µm

204

AC	anterior cerebellar tract	RV	rhombencephalic ventricle
CCe	corpus cerebelli	SR	superior raphe
Cven	commissura ventralis rhombencephali	SRF	superior reticular formation
DIL	diffuse nucleus of the inferior lobe	TeO	tectum opticum
DIV	trochlear decussation	TL	torus longitudinalis
GC	griseum centrale	TSc	central nucleus of torus semicircularis
LC	locus coeruleus	TSvl	ventrolateral nucleus of torus semicircularis
LLF	lateral longitudinal fascicle	TTB	tractus tectobulbaris
MLF	medial longitudinal fascicle	TTBc	tractus tectobulbaris cruciatus
NI	nucleus isthmi	TTBr	tractus tectobulbaris rectus
NLV	nucleus lateralis valvulae	Val	lateral division of valvula cerebelli
PGZ	periventricular gray zone of optic tectum	Vam	medial division of valvula cerebelli

Cross Section 208

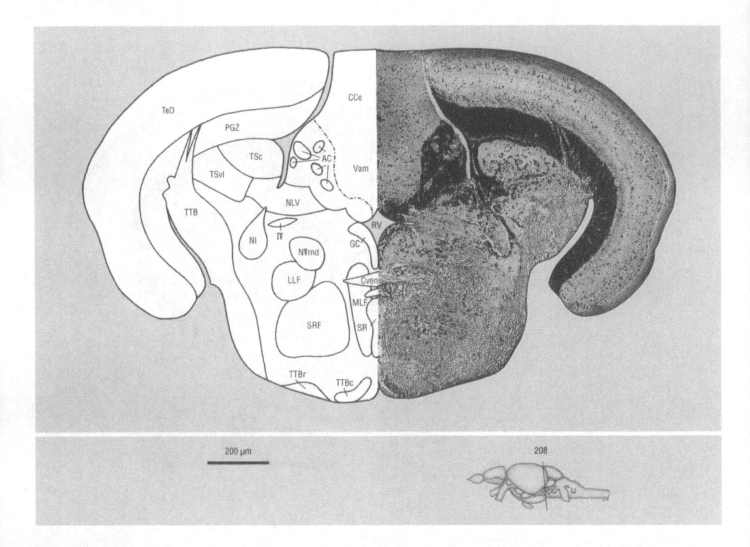

200 µm

208

AC	anterior cerebellar tract	SR	superior raphe
CCe	corpus cerebelli	SRF	superior reticular formation
Cven	commissura ventralis rhombencephali	TeO	tectum opticum
GC	griseum centrale	TSc	central nucleus of torus semicircularis
LLF	lateral longitudinal fascicle	TSvl	ventrolateral nucleus of torus semicircularis
MLF	medial longitudinal fascicle	TTB	tractus tectobulbaris
NI	nucleus isthmi	TTBc	tractus tectobulbaris cruciatus
NLV	nucleus lateralis valvulae	TTBr	tractus tectobulbaris rectus
NⅤmd	trigeminal motor nucleus, dorsal part	Vam	medial division of valvula cerebelli
PGZ	periventricular gray zone of optic tectum	Ⅳ	trochlear nerve
RV	rhombencephalic ventricle		

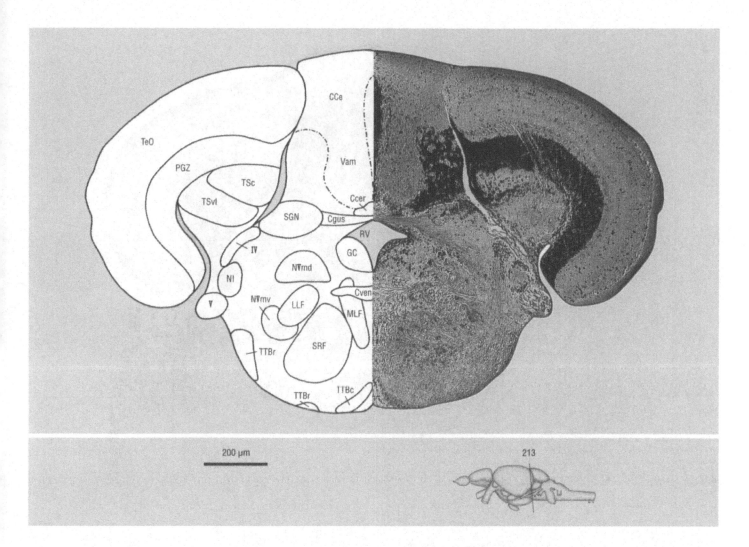

200 µm

213

CCe	corpus cerebelli
Ccer	commissura cerebelli
Cgus	commissure of the secondary gustatory nuclei
Cven	commissura ventralis rhombencephali
GC	griseum centrale
LLF	lateral longitudinal fascicle
MLF	medial longitudinal fascicle
NI	nucleus isthmi
NⅤmd	trigeminal motor nucleus, dorsal part
NⅤmv	trigeminal motor nucleus, ventral part
PGZ	periventricular gray zone of optic tectum
RV	rhombencephalic ventricle

SGN	secondary gustatory nucleus
SRF	superior reticular formation
TeO	tectum opticum
TSc	central nucleus of torus semicircularis
TSvl	ventrolateral nucleus of torus semicircularis
TTBc	tractus tectobulbaris cruciatus
TTBr	tractus tectobulbaris rectus
Vam	medial division of valvula cerebelli
Ⅳ	trochlear nerve
Ⅴ	trigeminal nerve

Cross Section 219

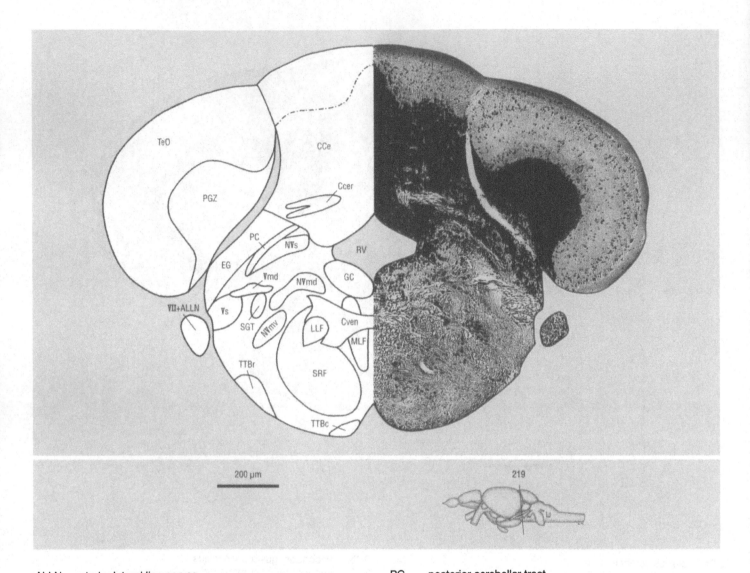

200 µm

219

ALLN	anterior lateral line nerves	PC	posterior cerebellar tract
CCe	corpus cerebelli	PGZ	periventricular gray zone of optic tectum
Ccer	commissura cerebelli	RV	rhombencephalic ventricle
Cven	commissura ventralis rhombencephali	SGT	secondary gustatory tract
EG	eminentia granularis	SRF	superior reticular formation
GC	griseum centrale	TeO	tectum opticum
LLF	lateral longitudinal fascicle	TTBc	tractus tectobulbaris cruciatus
MLF	medial longitudinal fascicle	TTBr	tractus tectobulbaris rectus
NⅤmd	trigeminal motor nucleus, dorsal part	Ⅴmd	dorsal motor root of the trigeminal nerve
NⅤmv	trigeminal motor nucleus, ventral part	Ⅴs	sensory root of the trigeminal nerve
NⅤs	primary sensory trigeminal nucleus	Ⅶ	facial nerve

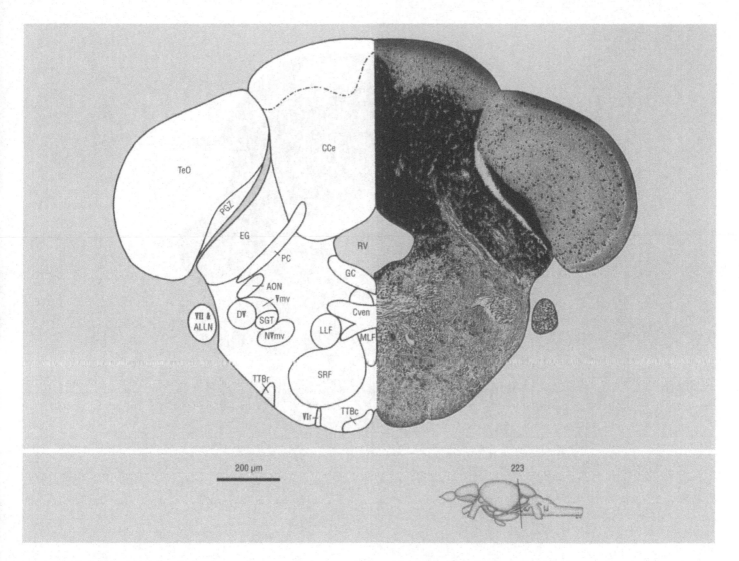

200 µm

223

ALLN	anterior lateral line nerves	PGZ	periventricular gray zone of optic tectum
AON	anterior octaval nucleus	RV	rhombencephalic ventricle
CCe	corpus cerebelli	SGT	secondary gustatory tract
Cven	commissura ventralis rhombencephali	SRF	superior reticular formation
DV	descending trigeminal root	TeO	tectum opticum
EG	eminentia granularis	TTBc	tractus tectobulbaris cruciatus
GC	griseum centrale	TTBr	tractus tectobulbaris rectus
LLF	lateral longitudinal fascicle	Vmv	ventral motor root of the trigeminal nerve
MLF	medial longitudinal fascicle	VIr	rostral root of the abducens nerve
NVmv	trigeminal motor nucleus, ventral part	VII	facial nerve
PC	posterior cerebellar tract		

Cross Section 230

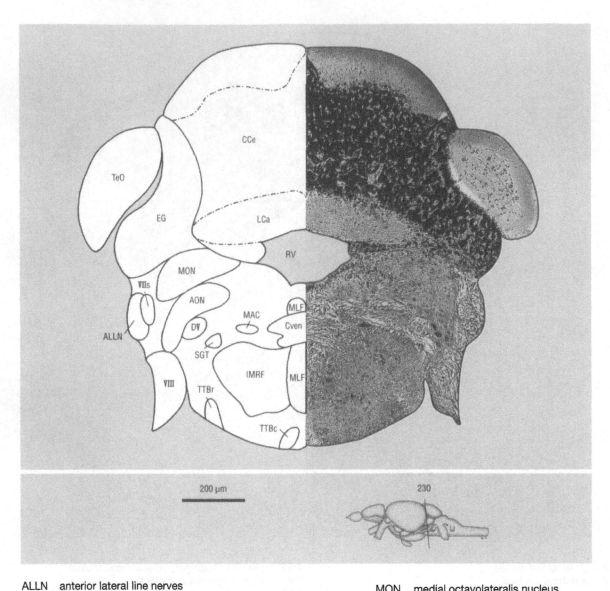

200 µm

230

ALLN	anterior lateral line nerves	MON	medial octavolateralis nucleus
AON	anterior octaval nucleus	RV	rhombencephalic ventricle
CCe	corpus cerebelli	SGT	secondary gustatory tract
Cven	commissura ventralis rhombencephali	TeO	tectum opticum
DV	descending trigeminal root	TTBc	tractus tectobulbaris cruciatus
EG	eminentia granularis	TTBr	tractus tectobulbaris rectus
IMRF	intermediate reticular formation		
LCa	lobus caudalis cerebelli	VIIs	sensory root of the facial nerve
MAC	Mauthner cell	VIII	octaval nerve
MLF	medial longitudinal fascicle		

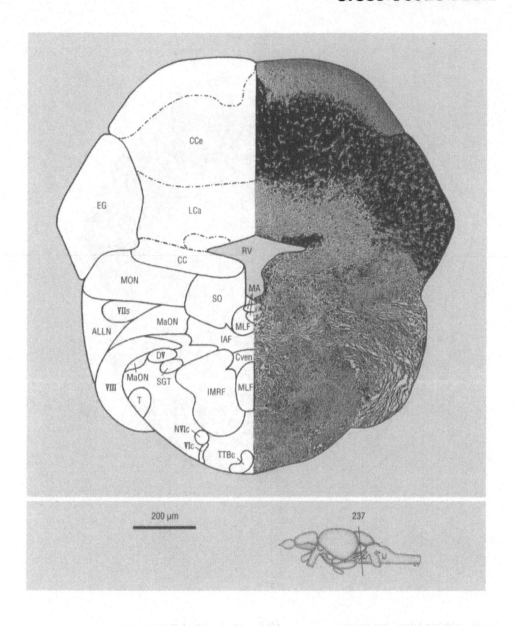

200 µm

237

ALLN	anterior lateral line nerves	MON	medial octavolateralis nucleus
CC	crista cerebellaris	NVIc	abducens nucleus, caudal part
CCe	corpus cerebelli	RV	rhombencephalic ventricle
Cven	commissura ventralis rhombencephali	SGT	secondary gustatory tract
DV	descending trigeminal root	SO	secondary octaval population (of McCormick & Hernandez 95)
EG	eminentia granularis		
IAF	inner arcuate fibers	T	tangential nucleus
IMRF	intermediate reticular formation	TTBc	tractus tectobulbaris cruciatus
LCa	lobus caudalis cerebelli		
MA	Mauthner axon	VIc	caudal root of the abducens nerve
MaON	magnocellular octaval nucleus	VIIs	sensory root of the facial nerve
MLF	medial longitudinal fascicle	VIII	octaval nerve

Cross Section 239

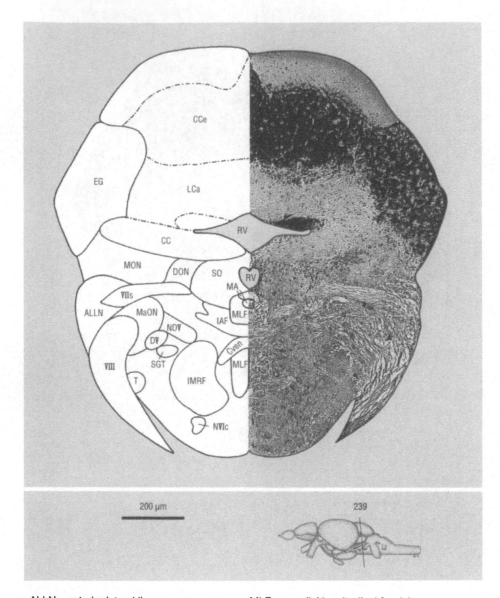

200 μm

239

ALLN	anterior lateral line nerves	MLF	medial longitudinal fascicle
CC	crista cerebellaris	MON	medial octavolateralis nucleus
CCe	corpus cerebelli	NDV	nucleus of the descending trigeminal root
Cven	commissura ventralis rhombencephali		
DON	descending octaval nucleus	NVIc	abducens nucleus, caudal part
DV	descending trigeminal root	RV	rhombencephalic ventricle
EG	eminentia granularis	SGT	secondary gustatory tract
IAF	inner arcuate fibers	SO	secondary octaval population (of McCormick & Hernandez 95)
IMRF	intermediate reticular formation		
LCa	lobus caudalis cerebelli	T	tangential nucleus
MA	Mauthner axon		
MaON	magnocellular octaval nucleus	VIIs	sensory root of the facial nerve
		VIII	octaval nerve

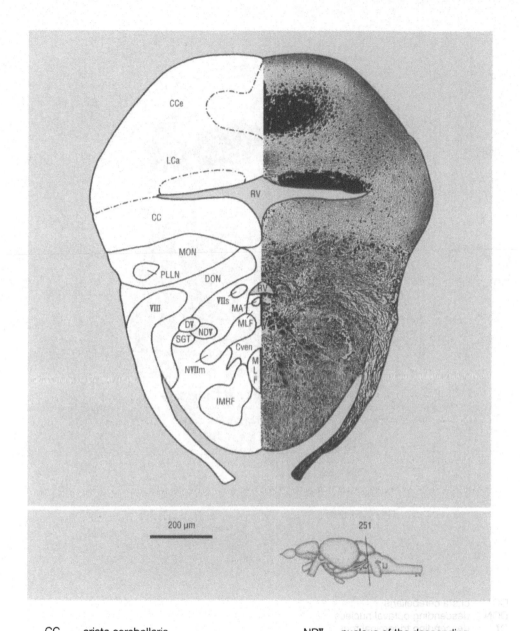

200 µm

251

CC	crista cerebellaris	NDV	nucleus of the descending trigeminal root
CCe	corpus cerebelli		
Cven	commissura ventralis rhombencephali	NVIIm	facial motor nucleus
DON	descending octaval nucleus	PLLN	posterior lateral line nerve
DV	descending trigeminal root	RV	rhombencephalic ventricle
IMRF	intermediate reticular formation	SGT	secondary gustatory tract
LCa	lobus caudalis cerebelli		
MA	Mauthner axon	VIIs	sensory root of the facial nerve
MLF	medial longitudinal fascicle	VIII	octaval nerve
MON	medial octavolateralis nucleus		

Cross Section 260

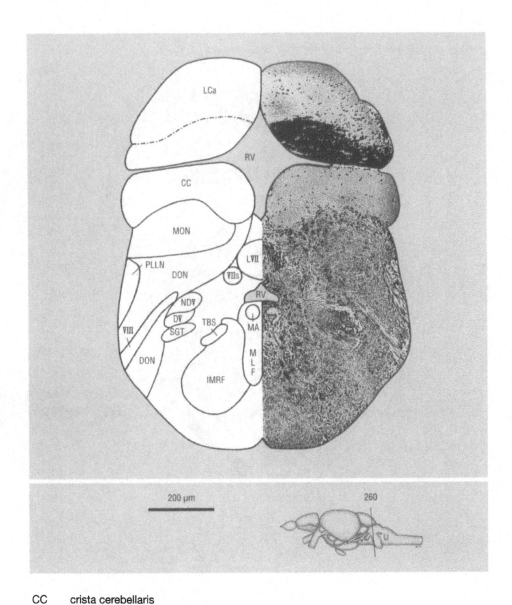

200 µm

260

CC   crista cerebellaris
DON   descending octaval nucleus
DV   descending trigeminal root
IMRF   intermediate reticular formation
LCa   lobus caudalis cerebelli
LVII   lobus facialis
MA   Mauthner axon
MLF   medial longitudinal fascicle
MON   medial octavolateralis nucleus
NDV   nucleus of the descending trigeminal root
PLLN   posterior lateral line nerve
RV   rhombencephalic ventricle
SGT   secondary gustatory tract
TBS   tractus bulbospinalis

VIIs   sensory root of the facial nerve
VIII   octaval nerve

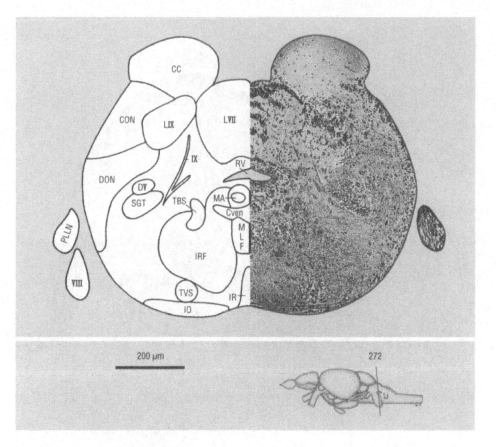

200 µm

272

CC	crista cerebellaris
CON	caudal octavolateralis nucleus
Cven	commissura ventralis rhombencephali
DON	descending octaval nucleus
DV	descending trigeminal root
IO	oliva inferior
IR	interior raphe
IRF	inferior reticular formation
LVII	lobus facialis
LIX	lobus glossopharyngeus
MA	Mauthner axon
MLF	medial longitudinal fascicle
PLLN	posterior lateral line nerve
RV	rhombencephalic ventricle
SGT	secondary gustatory tract
TBS	tractus bulbospinalis
TVS	tractus vestibulospinalis
VIII	octaval nerve
IX	glossopharyngeal nerve

Cross Section 279

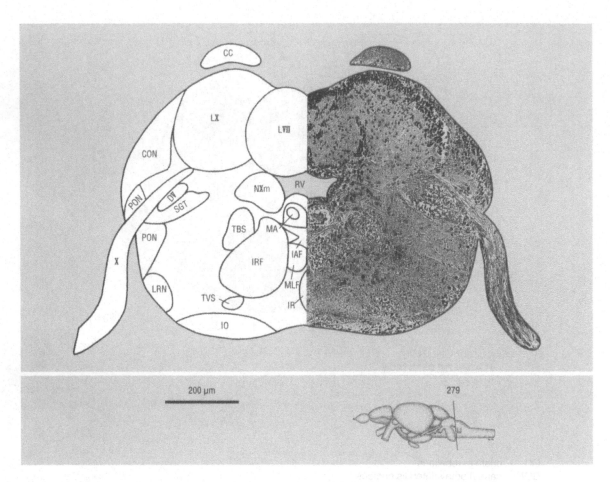

200 µm

279

CC	crista cerebellaris
CON	caudal octavolateralis nucleus
DV	descending trigeminal root
IAF	inner arcuate fibers
IO	oliva inferior
IR	inferior raphe
IRF	inferior reticular formation
LRN	lateral reticular nucleus
LVII	lobus facialis
LX	lobus vagus
MA	Mauthner axon
MLF	medial longitudinal fascicle
NXm	vagal motor nucleus
PON	posterior octaval nucleus
RV	rhombencephalic ventricle
SGT	secondary gustatory tract
TBS	tractus bulbospinalis
TVS	tractus vestibulospinalis
X	vagal nerve

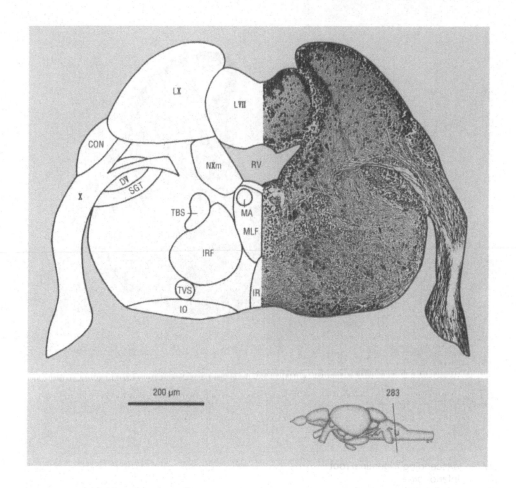

200 µm

283

CON	caudal octavolateralis nucleus
DV	descending trigeminal root
IO	inferior olive
IR	inferior raphe
IRF	inferior reticular formation
LVII	lobus facialis
LX	lobus vagus
MA	Mauthner axon
MLF	medial longitudinal fascicle
NXm	vagal motor nucleus
RV	rhombencephalic ventricle
SGT	secondary gustatory tract
TBS	tractus bulbospinalis
TVS	tractus vestibulospinalis
X	vagal nerve

Cross Section 290

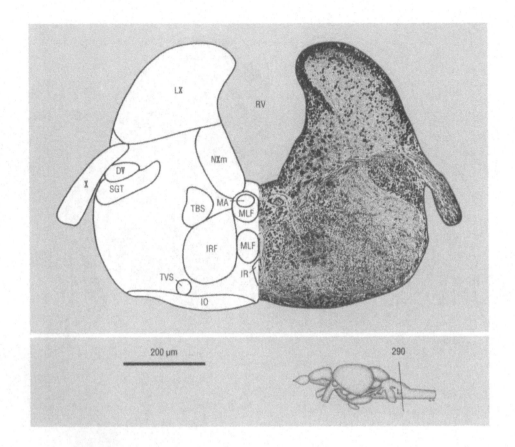

200 µm

290

DV	descending trigeminal root
IO	inferior olive
IR	inferior raphe
IRF	inferior reticular formation
LX	lobus vagus
MA	Mauthner axon
MLF	medial longitudinal fascicle
NXm	vagal motor nucleus
RV	rhombencephalic ventricle
SGT	secondary gustatory tract
TBS	tractus bulbospinalis
TVS	tractus vestibulospinalis
X	vagal nerve

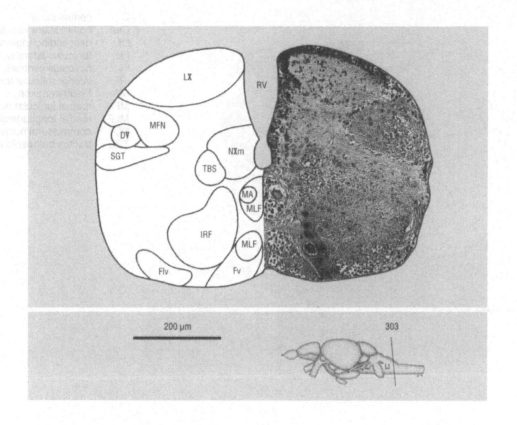

200 µm

303

DV	descending trigeminal root
Flv	funiculus lateralis pars ventralis
Fv	funiculus ventralis
IRF	inferior reticular formation
LX	lobus vagus
MA	Mauthner axon
MFN	medial funicular nucleus
MLF	medial longitudinal fascicle
NXm	vagal motor nucleus
RV	rhombencephalic ventricle
SGT	secondary gustatory tract
TBS	tractus bulbospinalis

Cross Section 319

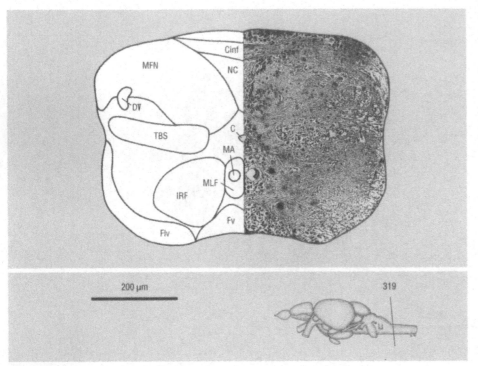

C	central canal
Cinf	commissura infima of Haller
DⱯ	descending trigeminal root
Flv	funiculus lateralis pars ventralis
Fv	funiculus ventralis
IRF	inferior reticular formation
MA	Mauthner axon
MFN	medial funicular nucleus
MLF	medial longitudinal fascicle
NC	commissural nucleus of Cajal
TBS	tractus bulbospinalis

200 µm

319

Cross Section 363

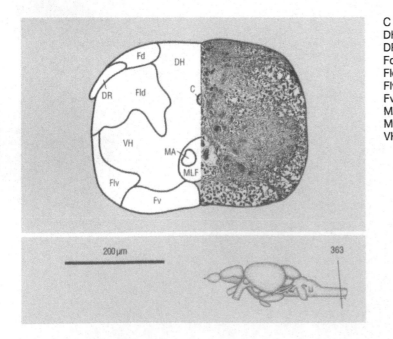

C	central canal
DH	dorsal horn
DR	dorsal root
Fd	funiculus dorsalis
Fld	funiculus lateralis pars dorsalis (including TBS)
Flv	funiculus lateralis pars ventralis
Fv	funiculus ventralis (including part of MLF)
MA	Mauthner axon
MLF	medial longitudinal fascicle
VH	ventral horn

200 µm

363

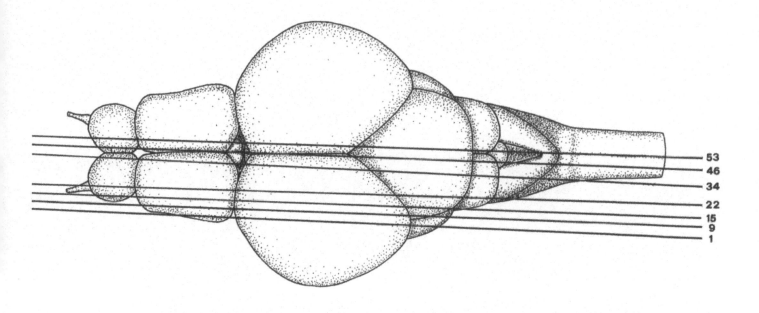

Figure 6.
Dorsal view of the adult zebrafish brain illustrating the position of levels shown in the series of sagittal sections. The series of sagittal sections shown on the following pages proceeds at irregular intervals from lateral to medial, giving an overview of the topological organization of the left half of the adult zebrafish brain. However, as the section plane does not correspond to the ideal longitudinal axis but deviates to the right side at rostral levels and to the left at caudal levels, neither section 46 nor section 53 represents a perfect midsagittal section.

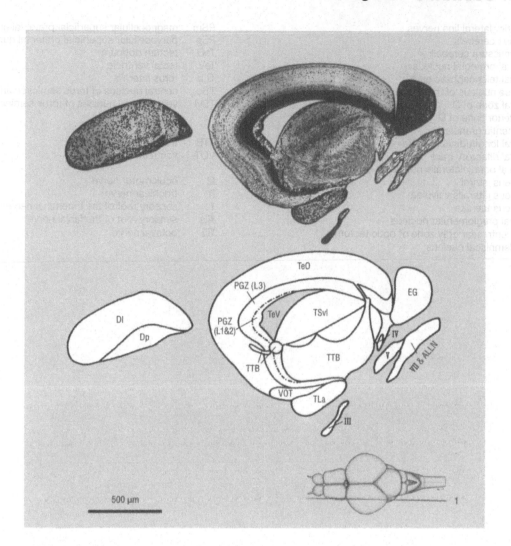

ALLN   anterior lateral line nerves
D   dorsal telencephalic area
Dl   lateral zone of D
Dp   posterior zone of D
EG   eminentia granularis
PGZ   periventricular gray zone of optic tectum
TeO   tectum opticum
TeV   tectal ventricle
TLa   torus lateralis
TSvl   ventrolateral nucleus of torus semicircularis
TTB   tractus tectobulbaris
VOT   ventrolateral optic tract

III   oculomotor nerve
IV   trochlear nerve
V   trigeminal nerve
VII   facial nerve

Sagittal Section 9

ALLN anterior lateral line nerves
CCe corpus cerebelli
Ccer commissura cerebelli
CPN central pretectal nucleus
D dorsal telencephalic area
DIL diffuse nucleus of the inferior lobe
Dl lateral zone of D
Dp posterior zone of D
EG eminentia granularis
LLF lateral longitudinal fascicle
LOT lateral olfactory tract
MON medial octavolateralis nucleus
NI nucleus isthmi
NLV nucleus lateralis valvulae
NT nucleus taeniae
PGl lateral preglomerular nucleus
PGZ periventricular gray zone of optic tectum
PL perilemniscal nucleus

PSm magnocellular superficial pretectal nucleus
PSp parvocellular superficial pretectal nucleus
TeO tectum opticum
TeV tectal ventricle
TLa torus lateralis
TSc central nucleus of torus semicircularis
TSvl ventrolateral nucleus of torus semicircularis

TTB tractus tectobulbaris
VOT ventrolateral optic tract

III oculomotor nerve
IV trochlear nerve
Vs sensory root of the trigeminal nerve
VIIs sensory root of the facial nerve
VIII octaval nerve

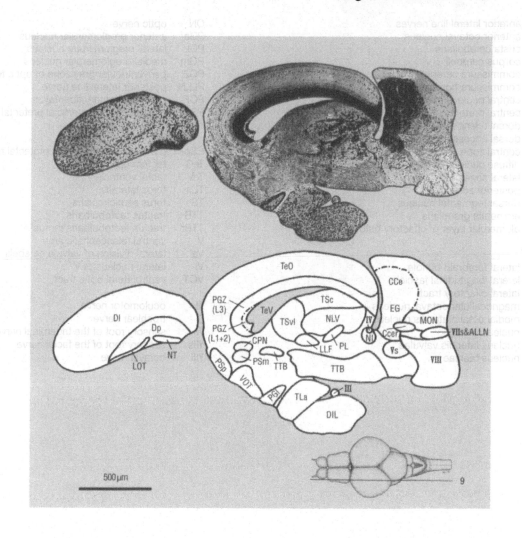

500µm

9

Sagittal Section 15

ALLN	anterior lateral line nerves		ON	optic nerve
AON	anterior octaval nucleus		PGa	anterior preglomerular nucleus
CC	crista cerebellaris		PGl	lateral preglomerular nucleus
CCe	corpus cerebelli		PGm	medial preglomerular nucleus
Ccer	commissura cerebelli		PGZ	periventricular gray zone of optic tectum
Chor	commissura horizontalis		PLLN	posterior lateral line nerve
CIL	central nucleus of the inferior lobe		POF	primary olfactory fiber layer
CPN	central pretectal nucleus		PSm	magnocellular superficial pretectal nucleus
D	dorsal telencephalic area			
DAO	dorsal accessory optic nucleus			
Dc	central zone of D		PSp	parvocellular superficial pretectal nucleus
DIL	diffuse nucleus of the inferior lobe		TeO	tectum opticum
Dl	lateral zone of D		TeV	tectal ventricle
Dp	posterior zone of D		TLa	torus lateralis
DTN	dorsal tegmental nucleus		TS	torus semicircularis
EG	eminentia granularis		TTB	tractus tectobulbaris
GL	glomerular layer of olfactory bulb		TTBr	tractus tectobulbaris rectus
			V	ventral telencephalic area
			Val	lateral division of valvula cerebelli
LFB	lateral forebrain bundle		Vl	lateral nucleus of V
LLF	lateral longitudinal fascicle		VOT	ventrolateral optic tract
LOT	lateral olfactory tract			
MaON	magnocellular octaval nucleus		III	oculomotor nerve
MON	medial octavolateralis nucleus		IV	trochlear nerve
NI	nucleus isthmi		Vs	sensory root of the trigeminal nerve
NLV	nucleus lateralis valvulae		VIIs	sensory root of the facial nerve
NT	nucleus taeniae		VIII	octaval nerve

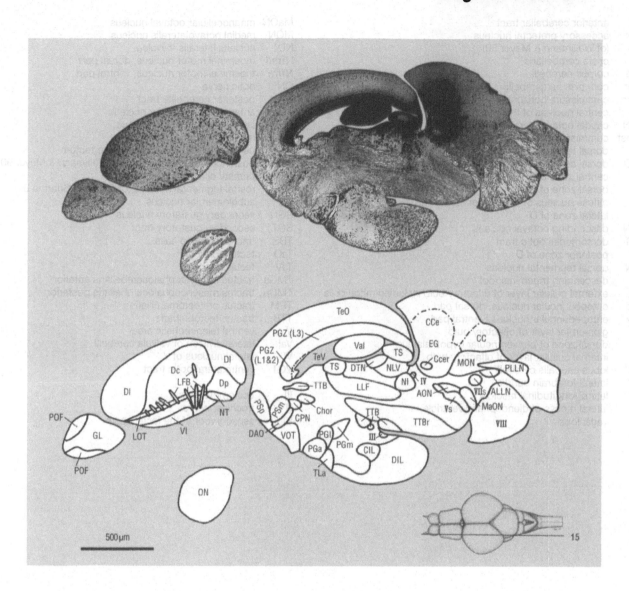

500μm

15

AC	anterior cerebellar tract	MaON	magnocellular octaval nucleus
APN	accessory pretectal nucleus	MON	medial octavolateralis nucleus
	(of Wullimann & Meyer 90)	NLV	nucleus lateralis valvulae
CC	crista cerebellaris	NⅤmd	trigeminal motor nucleus, dorsal part
CCe	corpus cerebelli	NⅤmv	trigeminal motor nucleus, ventral part
Ccer	commissura cerebelli	ON	optic nerve
Chor	commissura horizontalis	PC	posterior cerebellar tract
CIL	central nucleus of the inferior lobe	PGa	anterior preglomerular nucleus
CON	caudal octavolateralis nucleus	PGl	lateral preglomerular nucleus
Cpost	commissura posterior	PGm	medial preglomerular nucleus
D	dorsal telencephalic area	PGZ	periventricular grey zone of optic tectum
DAO	dorsal accessory optic nucleus	PO	posterior pretectal nucleus (of Wullimann & Meyer 90)
Dc	central zone of D	POF	primary olfactory fiber layer
Dd	dorsal zone of D	RT	rostral tegmental nucleus (of Grover & Sharma 81)
DIL	diffuse nucleus of the inferior lobe	SG	subglomerular nucleus
Dl	lateral zone of D	SGN	secondary gustatory nucleus
DON	descending octaval nucleus	SGT	secondary gustatory tract
DOT	dorsomedial optic tract	TBS	tractus bulbospinalis
Dp	posterior zone of D	TeO	tectum opticum
DTN	dorsal tegmental nucleus	TeV	tectal ventricle
DⅤ	descending trigeminal root	TMCa	tractus mesencephalocerebellaris anterior
ECL	external cellular layer of olfactory bulb including mitral cells	TMCp	tractus mesencephalocerebellaris posterior
ENd	entopeduncular nucleus, dorsal part	TPM	tractus pretectomamillaris
ENv	entopeduncular nucleus, ventral part	TTB	tractus tectobulbaris
GL	glomerular layer of olfactory bulb	V	ventral telencephalic area
Hd	dorsal zone of periventricular hypothalamus	Val	lateral division of valvula cerebelli
ICL	internal cellular layer of olfactory bulb	Vl	lateral nucleus of V
LCa	lobus caudalis cerebelli	VOT	ventrolateral optic tract
LFB	lateral forebrain bundle		
LLF	lateral longitudinal fascicle	Ⅲ	oculomotor nerve
LR	lateral recess of diencephalic ventricle	Ⅳ	trochlear nerve
LX	vagal lobe	VIIs	sensory root of the facial nerve

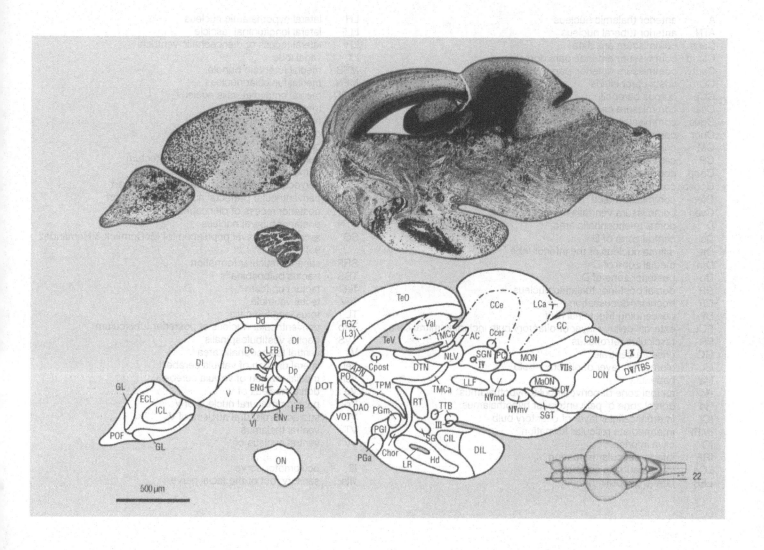

500 µm

Sagittal Section 34

A	anterior thalamic nucleus	LH	lateral hypothalamic nucleus
ATN	anterior tuberal nucleus	LLF	lateral longitudinal fascicle
Cans	commissura ansulata	LR	lateral recess of diencephalic ventricle
Cantd	commissura anterior, pars dorsalis	LX	vagal lobe
Cantv	commissura anterior, pars ventralis	MFB	medial forebrain bundle
CC	crista cerebellaris	MFN	medial funicular nucleus
CCe	corpus cerebelli	MON	medial octavolateralis nucleus
Ccer	commissura cerebelli	MOT	medial olfactory tract
Cgus	commissure of the secondary gustatory nuclei	NLV	nucleus lateralis valvulae
Chor	commissura horizontalis	ON	optic nerve
CM	corpus mamillare	OT	optic tract
CP	central posterior thalamic nucleus	PGZ	periventricular gray zone of optic tectum
Cpop	commissura postoptica	POF	primary olfactory fiber layer
Cpost	commissura posterior	PPa	parvocellular preoptic nucleus, anterior part
Ctec	commissura tecti	PPv	periventricular pretectal nucleus, ventral part
Cven	commissura ventralis rhombencephali	PR	posterior recess of diencephalic ventricle
D	dorsal telencephalic area	PTN	posterior tuberal nucleus
Dc	central zone of D	SO	secondary octaval population (of McCormick & Hernandez 95)
DIL	diffuse nucleus of the inferior lobe		
Dm	medial zone of D	SRF	superior reticular formation
Dp	posterior zone of D	TBS	tractus bulbospinalis
DP	dorsal posterior thalamic nucleus	TeO	tectum opticum
DIV	trochlear decussation	TeV	tectal ventricle
DV	descending trigeminal root	TL	torus longitudinalis
ECL	external cellular layer of olfactory bulb including mitral cells	TPp	periventricular nucleus of posterior tuberculum
FR	fasciculus retroflexus	TVS	tractus vestibulospinalis
GC	griseum centrale	V	ventral telencephalic area
GL	glomerular layer of olfactory bulb	Val	lateral division of valvula cerebelli
Ha	habenula	Vam	medial division of valvula cerebelli
Hc	caudal zone of periventricular hypothalamus	Vd	dorsal nucleus of V
Hd	dorsal zone of periventricular hypothalamus	Vp	postcommissural nucleus of V
ICL	internal cellular layer of olfactory bulb	Vs	supracommissural nucleus of V
IMRF	intermediate reticular formation	VT	ventral thalamus
IO	oliva inferior	Vv	ventral nucleus of V
IRF	inferior reticular formation		
LCa	lobus caudalis cerebelli	III	oculomotor nerve
LFB	lateral forebrain bundle	VIIs	sensory root of the facial nerve

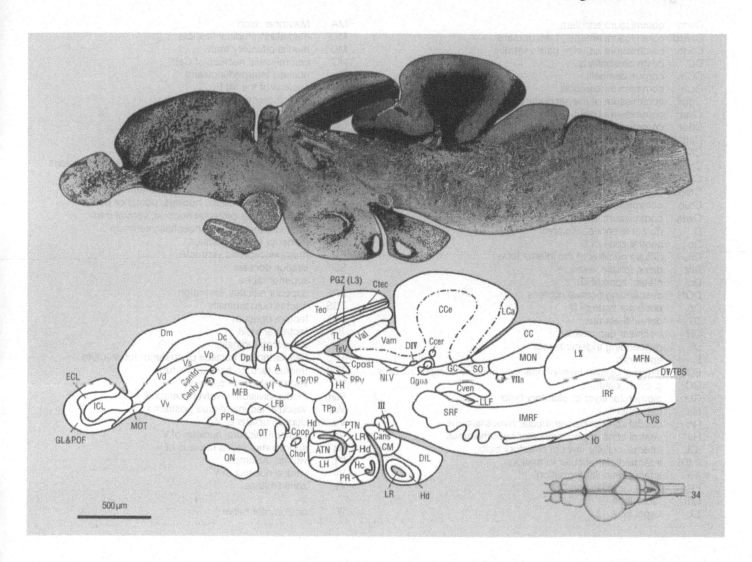

Sagittal Section 46

Cans	commissura ansulata	MA	Mauthner axon
Cantd	commissura anterior, pars dorsalis	MLF	medial longitudinal fascicle
Cantv	commissura anterior, pars ventralis	MOT	medial olfactory tract
CC	crista cerebellaris	NC	commissural nucleus of Cajal
CCe	corpus cerebelli	NIn	nucleus interpeduncularis
Ccer	commissura cerebelli	NMLF	nucleus of the MLF
Cgus	commissure of the secondary gustatory nuclei	NIII	oculomotor nucleus
Chab	commissura habenularum	NXm	vagal motor nucleus
Chor	commissura horizontalis	ON	optic nerve
Cinf	commissura infima of Haller	PGZ	periventricular gray zone of optic tectum
CM	corpus mamillare	PM	magnocellular preoptic nucleus
Cpop	commissura postoptica	PMg	gigantocellular part of magnocellular preoptic nucleus
Cpost	commissura posterior	POF	primary olfactory fiber layer
Ctec	commissura tecti	PPa	parvocellular preoptic nucleus, anterior part
Ctub	commissure of the posterior tuberculum	PPp	parvocellular preoptic nucleus, posterior part
Cven	commissura ventralis rhombencephali	PPv	periventricular pretectal nucleus, ventral part
D	dorsal telencephalic area	PR	posterior recess of diencephalic ventricle
Dc	central zone of D	PTN	posterior tuberal nucleus
DIL	diffuse nucleus of the inferior lobe	RV	rhombencephalic ventricle
DiV	diencephalic ventricle	SD	saccus dorsalis
Dm	medial zone of D	SR	superior raphe
DON	descending octaval nucleus	SRF	superior reticular formation
Dp	posterior zone of D	TBS	tractus bulbospinalis
DT	dorsal thalamus	TeO	tectum opticum
DIV	trochlear decussation	TeV	tectal ventricle
DV	descending trigeminal root	TL	torus longitudinalis
E	epiphysis	TPp	periventricular nucleus of posterior tuberculum
ECL	external cellular layer of olfactory bulb including mitral cells	TTBc	tractus tectobulbaris cruciatus
GC	griseum centrale	V	ventral telencephalic area
GL	glomerular layer of olfactory bulb	Vam	medial division of valvula cerebelli
Ha	habenula	Vas	vascular lacuna of area postrema
Hc	caudal zone of periventricular hypothalamus	Vd	dorsal nucleus of V
Hv	ventral zone of periventricular hypothalamus	Vp	postcommissural nucleus of V
ICL	internal cellular layer of olfactory bulb	Vs	supracommissural nucleus of V
IMRF	intermediate reticular formation	VT	ventral thalamus
IRF	inferior reticular formation	Vv	ventral nucleus of V
LCa	lobus caudalis cerebelli	ZL	zona limitans
LVII	facial lobe		
LX	vagal lobe	III	oculomotor nerve

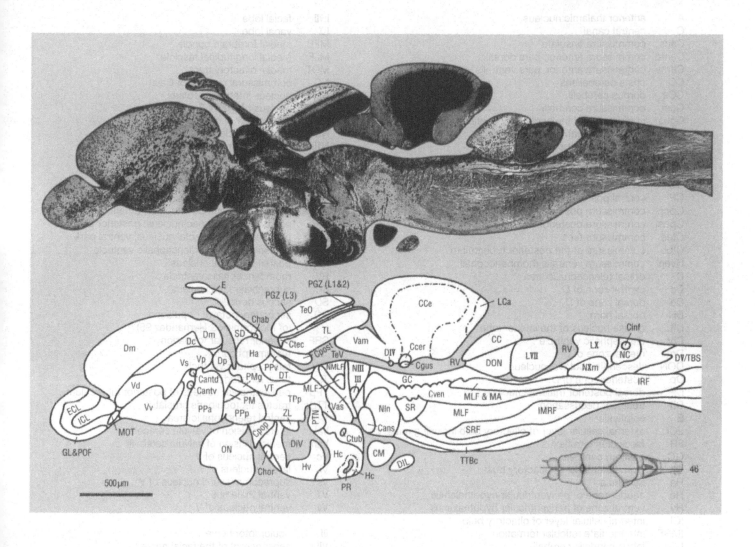

Sagittal Section 53

A	anterior thalamic nucleus
C	central canal
Cans	commissura ansulata
Cantd	commissura anterior, pars dorsalis
Cantv	commissura anterior, pars ventralis
CC	crista cerebellaris
CCe	corpus cerebelli
Ccer	commissura cerebelli
Cgus	commissure of the secondary gustatory nuclei
Chab	commissura habenularum
Chor	commissura horizontalis
Cinf	commissura infima of Haller
CM	corpus mamillare
CO	chiasma opticum
CP	central posterior thalamic nucleus
Cpop	commissura postoptica
Cpost	commissura posterior
Ctec	commissura tecti
Ctub	commissure of the posterior tuberculum
Cven	commissura ventralis rhombencephali
D	dorsal telencephalic area
Dc	central zone of D
Dd	dorsal zone of D
DH	dorsal horn
DIL	diffuse nucleus of the inferior lobe
DiV	diencephalic ventricle
Dm	medial zone of D
DON	descending octaval nucleus
Dp	posterior zone of D
DP	dorsal posterior thalamic nucleus
DIV	trochlear decussation
E	epiphysis
ECL	external cellular layer of olfactory bulb including mitral cells
FR	fasciculus retroflexus
GC	griseum centrale
GL	glomerular layer of olfactory bulb
Ha	habenula
Hc	caudal zone of periventricular hypothalamus
Hv	ventral zone of periventricular hypothalamus
ICL	internal cellular layer of olfactory bulb
IMRF	intermediate reticular formation
LCa	lobus caudalis cerebelli

LVII	facial lobe
LX	vagal lobe
MFB	medial forebrain bundle
MLF	medial longitudinal fascicle
MOT	medial olfactory tract
NC	commissural nucleus of Cajal
NIn	nucleus interpeduncularis
NMLF	nucleus of the MLF
NIII	oculomotor nucleus
NIV	trochlear nucleus
NVIIm	facial motor nucleus
NXm	vagal motor nucleus
ON	optic nerve
PGZ	periventricular gray zone of optic tectum
POF	primary olfactory fiber layer
PPa	parvocellular preoptic nucleus, anterior part
PPp	parvocellular preoptic nucleus, posterior part
PPv	periventricular pretectal nucleus, ventral part
PR	posterior recess of diencephalic ventricle
PTN	posterior tuberal nucleus
RV	rhombencephalic ventricle
SC	suprachiasmatic nucleus
SD	saccus dorsalis
SO	secondary octaval population (of McCormick & Hernandez 95)
SRF	superior reticular formation
TeO	tectum opticum
TeV	telencephalic ventricle
TL	torus longitudinalis
TPp	periventricular nucleus of posterior tuberculum
TTBc	tractus tectobulbaris cruciatus
V	ventral telencephalic area
Val	lateral division of valvula cerebelli
Vam	medial division of valvula cerebelli
Vc	central nucleus of V
Vl	lateral nucleus of V
Vs	supracommissural nucleus of V
VT	ventral thalamus
Vv	ventral nucleus of V
III	oculomotor nerve
VIIs	sensory root of the facial nerve

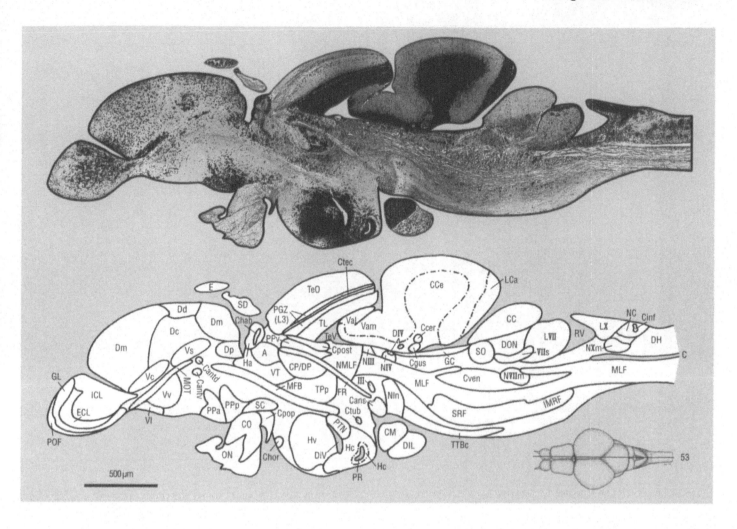

500 µm

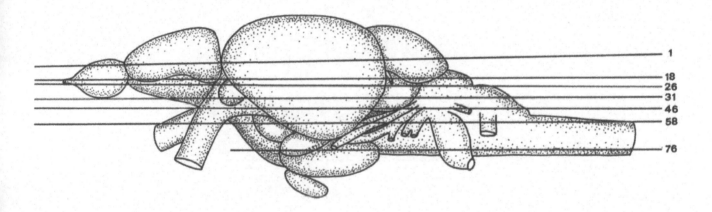

1
18
26
31
46
58
76

Figure 7.
Lateral view of the adult zebrafish brain demonstrating the position of levels shown in the series of horizontal sections. The selection of horizontal sections shown on the following pages does not include the most dorsal and ventral aspects of the brain. With regard to the spinal cord, the section plane of this series almost corresponds to the ideal horizontal axis.

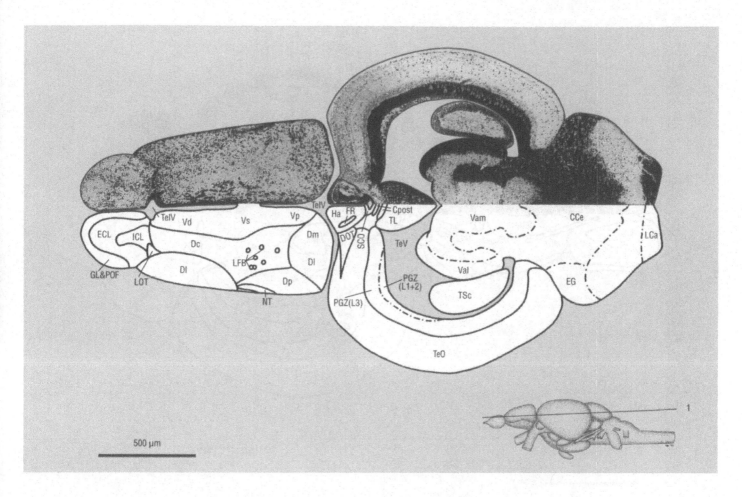

CCe	corpus cerebelli
Cpost	commissura posterior
D	dorsal telencephalic area
Dc	central zone of D
Dl	lateral zone of D
Dm	medial zone of D
DOT	dorsomedial optic tract
Dp	posterior zone of D
ECL	external cellular layer of olfactory bulb including mitral cells
EG	eminentia granularis
FR	fasciculus retroflexus
GL	glomerular layer of olfactory bulb
Ha	habenula
ICL	internal cellular layer of olfactory bulb
LCa	lobus caudalis cerebelli
LFB	lateral forebrain bundle
LOT	lateral olfactory tract
NT	nucleus taeniae

PGZ	periventricular gray zone of optic tectum
POF	primary olfactory fiber layer
SCO	subcommissural organ
TelV	telencephalic ventricles
TeO	tectum opticum
TeV	tectal ventricle
TL	torus longitudinalis
TSc	central nucleus of torus semicircularis
V	ventral telencephalic area
Val	lateral division of valvula cerebelli
Vam	medial division of valvula cerebelli
Vd	dorsal nucleus of V
Vp	postcommissural nucleus of V
Vs	supracommissural nucleus of V

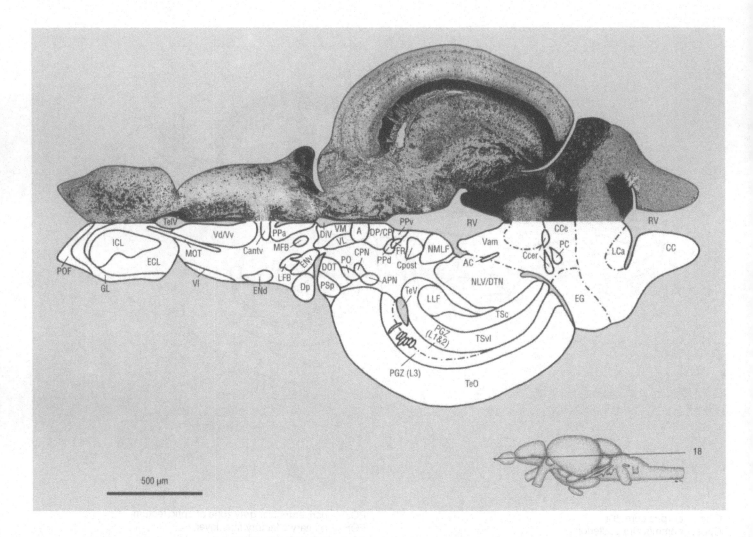

500 μm

A	anterior thalamic nucleus	LLF	lateral longitudinal fascicle
AC	anterior cerebellar tract	MFB	medial forebrain bundle
APN	accessory pretectal nucleus (of Wullimann & Meyer 90)	MOT	medial olfactory tract
Cantv	commissura anterior, ventral part	NLV	nucleus lateralis valvulae
CC	crista cerebellaris	NMLF	nucleus of the medial longitudinal fascicle
CCe	corpus cerebelli	PC	posterior cerebellar tract
Ccer	commissura cerebelli	PGZ	periventricular gray zone of optic tectum
CP	central posterior thalamic nucleus	PO	posterior pretectal nucleus (of Wullimann & Meyer 90)
CPN	central pretectal nucleus	POF	primary olfactory fiber layer
Cpost	commissura posterior	PPa	parvocellular preoptic nucleus, anterior part
D	dorsal telencephalic area	PPd	periventricular pretectal nucleus, dorsal part
DiV	diencephalic ventricle	PPv	periventricular pretectal nucleus, ventral part
DOT	dorsomedial optic tract	PSp	parvocellular superficial pretectal nucleus
Dp	posterior zone of D	RV	rhombencephalic ventricle
DP	dorsal posterior thalamic nucleus	TelV	telencephalic ventricles
DTN	dorsal tegmental nucleus	TeO	tectum opticum
ECL	external cellular layer of olfactory bulb including mitral cells	TeV	tectal ventricle
EG	eminentia granularis	TSc	central nucleus of torus semicircularis
ENd	entopeduncular nucleus, dorsal part	TSvl	ventrolateral nucleus of torus semicircularis
ENv	entopeduncular nucleus, ventral part	V	ventral telencephalic area
FR	fasciculus retroflexus	Vam	medial division of valvula cerebelli
GL	glomerular layer of olfactory bulb	Vd	dorsal nucleus of V
ICL	internal cellular layer of olfactory bulb	Vl	lateral nucleus of V
LCa	lobus caudalis cerebelli	VL	ventrolateral thalamic nucleus
LFB	lateral forebrain bundle	VM	ventromedial thalamic nucleus
		Vv	ventral nucleus of V

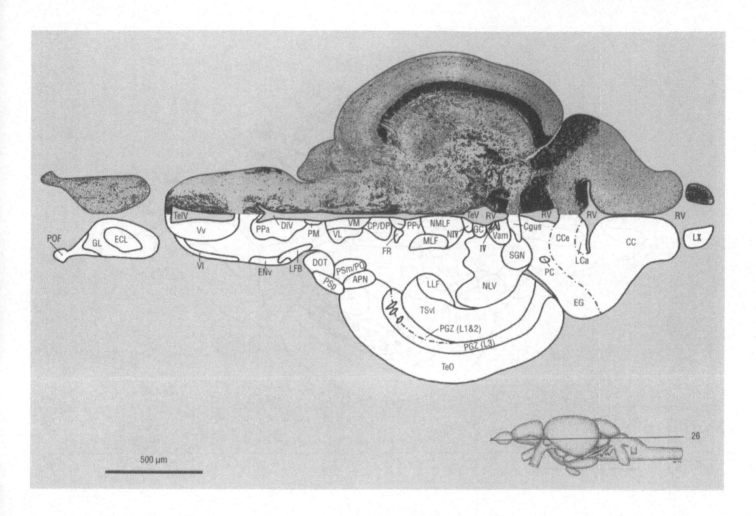

500 µm

APN	accessory pretectal nucleus (of Wullimann & Meyer 90)
CC	crista cerebellaris
CCe	corpus cerebelli
Cgus	commissure of the secondary gustatory nuclei
CP	central posterior thalamic nucleus
DiV	diencephalic ventricle
DOT	dorsomedial optic tract
DP	dorsal posterior thalamic nucleus
ECL	external cellular layer of olfactory bulb including mitral cells
EG	eminentia granularis
ENv	entopeduncular nucleus, ventral part
FR	fasciculus retroflexus
GC	griseum centrale
GL	glomerular layer of olfactory bulb
LCa	lobus caudalis cerebelli
LFB	lateral forebrain bundle
LLF	lateral longitudinal fascicle
LX	vagal lobe
MLF	medial longitudinal fascicle
NLV	nucleus lateralis valvulae
NMLF	nucleus of the medial longitudinal fascicle
NIV	trochlear nucleus
PC	posterior cerebellar tract

PGZ	periventricular gray zone of optic tectum
PM	magnocellular preoptic nucleus
PO	posterior pretectal nucleus (of Wullimann & Meyer 90)
POF	primary olfactory fiber layer
PPa	parvocellular preoptic nucleus, anterior part
PPv	periventricular pretectal nucleus, ventral part
PSm	magnocellular superficial pretectal nucleus
PSp	parvocellular superficial pretectal nucleus
RV	rhombencephalic ventricle
SGN	secondary gustatory nucleus
TelV	telencephalic ventricles
TeO	tectum opticum
TeV	tectal ventricle
TSvl	ventrolateral nucleus of torus semicircularis
V	ventral telencephalic area
Vam	medial division of valvula cerebelli
Vl	lateral nucleus of V
VL	ventrolateral thalamic nucleus
VM	ventromedial thalamic nucleus
Vv	ventral nucleus of V
IV	trochlear nerve

Horizontal Section 31

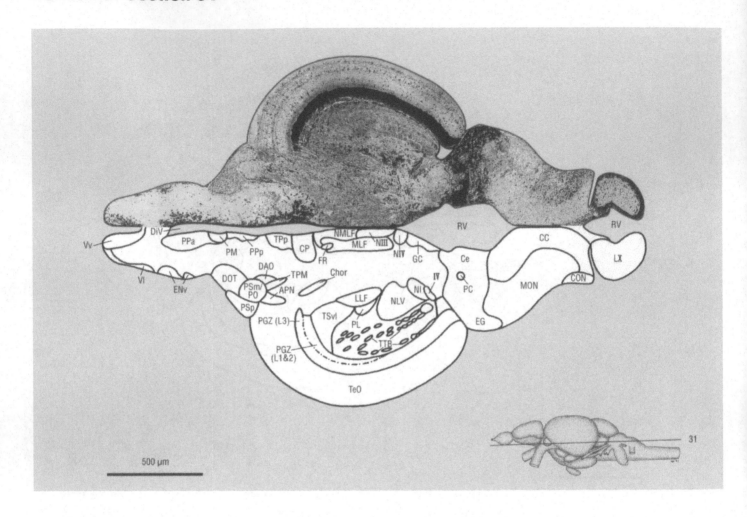

500 µm

31

APN	accessory pretectal nucleus (of Wullimann & Meyer 90)	NIV	trochlear nucleus
CC	crista cerebellaris	PC	posterior cerebellar tract
Ce	cerebellum	PGZ	periventricular gray zone of optic tectum
Chor	commissura horizontalis	PL	perilemniscal nucleus
CON	caudal octavolateralis nucleus	PM	magnocellular preoptic nucleus
CP	central posterior thalamic nucleus	PO	posterior pretectal nucleus (of Wullimann & Meyer 90)
DAO	dorsal accessory optic nucleus	PPa	parvocellular preoptic nucleus, anterior part
DiV	diencephalic ventricle	PPp	parvocellular preoptic nucleus, posterior part
DOT	dorsomedial optic tract	PSm	magnocellular superficial pretectal nucleus
EG	eminentia granularis	PSp	parvocellular superficial pretectal nucleus
ENv	entopeduncular nucleus, ventral part	RV	rhombencephalic ventricle
FR	fasciculus retroflexus	TeO	tectum opticum
GC	griseum centrale	TPM	tractus pretectomamillaris
LLF	lateral longitudinal fascicle	TPp	periventricular nucleus of posterior tuberculum
LX	vagal lobe	TSvl	nucleus ventrolateralis of torus semicircularis
MLF	medial longitudinal fascicle	TTB	tractus tectobulbaris
MON	medial octavolateralis nucleus	V	ventral telencephalic area
NI	nucleus isthmi	VI	lateral nucleus of V
NLV	nucleus lateralis valvulae	Vv	ventral nucleus of V
NMLF	nucleus of the medial longitudinal fascicle		
NIII	oculomotor nucleus	IV	trochlear nerve

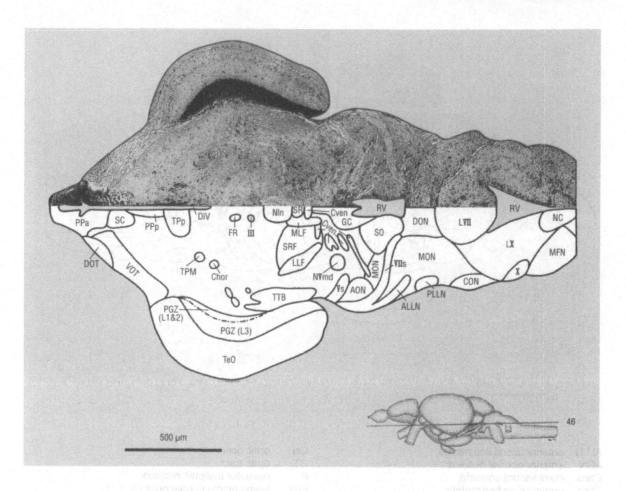

ALLN	anterior lateral line nerves	PLLN	posterior lateral line nerve
AON	anterior octaval nucleus	PPa	parvocellular preoptic nucleus, anterior part
Chor	commissura horizontalis	PPp	parvocellular preoptic nucleus, posterior part
CON	caudal octavolateralis nucleus	RV	rhombencephalic ventricle
Cven	commissura ventralis rhombencephali	SC	suprachiasmatic nucleus
DiV	diencephalic ventricle	SO	secondary octaval population
DON	descending octaval nucleus		(of McCormick & Hernandez 95)
DOT	dorsomedial optic tract	SR	superior raphe
FR	fasciculus retroflexus	SRF	superior reticular formation
GC	griseum centrale	TeO	tectum opticum
LLF	lateral longitudinal fascicle	TPM	tractus pretectomamillaris
LVII	facial lobe	TPp	periventricular nucleus of posterior tuberculum
LX	vagal lobe	TTB	tractus tectobulbaris
MFN	medial funicular nucleus	VOT	ventrolateral optic tract
MLF	medial longitudinal fascicle		
MON	medial octavolateralis nucleus	III	oculomotor nerve
NC	commissural nucleus of Cajal	Vs	sensory root of the trigeminal nerve
NIn	nucleus interpeduncularis	VIIs	sensory root of the facial nerve
NVmd	trigeminal motor nucleus, dorsal part	X	vagal nerve
PGZ	periventricular gray zone of optic tectum		

Horizontal Section 58

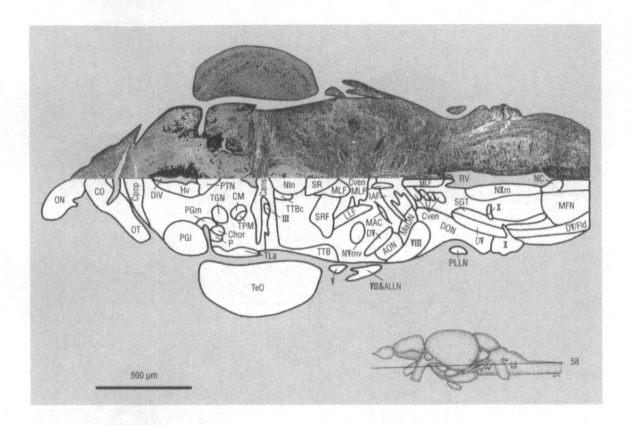

ALLN	anterior lateral line nerves	ON	optic nerve
AON	anterior octaval nucleus	OT	optic tract
Cans	commissura ansulata	P	posterior thalamic nucleus
Chor	commissura horizontalis	PGl	lateral preglomerular nucleus
CM	corpus mamillare	PGm	medial preglomerular nucleus
CO	chiasma opticum	PLLN	posterior lateral line nerve
Cpop	commissura postoptica	PTN	posterior tuberal nucleus
Cven	commissura ventralis rhombencephali	RV	rhombencephalic ventricle
DiV	diencephalic ventricle	SGT	secondary gustatory tract
DON	descending octaval nucleus	SR	superior raphe
DV	descending trigeminal root	SRF	superior reticular formation
Fld	funiculus lateralis, pars dorsalis	TeO	tectum opticum
Hv	ventral zone of periventricular hypothalamus	TGN	tertiary gustatory nucleus (of Wullimann 88)
IAF	inner arcuate fibers	TLa	torus lateralis
LLF	lateral longitudinal fascicle	TPM	tractus pretectomamillaris
MAC	Mauthner cell	TTB	tractus tectobulbaris
MaON	magnocellular octaval nucleus	TTBc	tractus tectobulbaris cruciatus
MFN	medial funicular nucleus		
MLF	medial longitudinal fascicle	III	oculomotor nerve
NC	commissural nucleus of Cajal	V	trigeminal nerve
NIn	nucleus interpeduncularis	VII	facial nerve
NVmv	trigeminal motor nucleus, ventral part	VIII	octaval nerve
NXm	vagal motor nucleus	X	vagal nerve

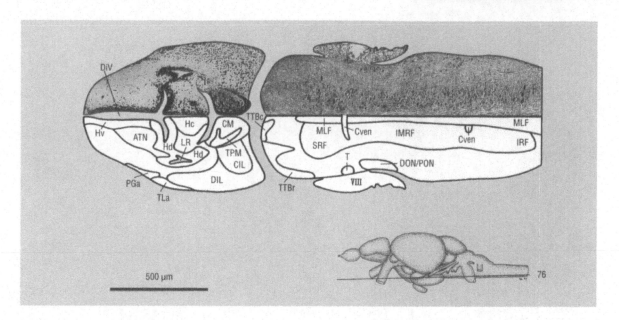

500 µm

76

ATN	anterior tuberal nucleus	LR	lateral recess of diencephalic ventricle
CIL	central nucleus of the inferior lobe	MLF	medial longitudinal fascicle
CM	corpus mamillare	PGa	anterior preglomerular nucleus
Cven	commissura ventralis rhombencephali	PON	posterior octaval nucleus
DIL	diffuse nucleus of the inferior lobe	SRF	superior reticular formation
DiV	diencephalic ventricle	TLa	torus lateralis
DON	descending octaval nucleus	TPM	tractus pretectomamillaris
Hc	caudal zone of periventricular hypothalamus	TTBc	tractus tectobulbaris cruciatus
Hd	dorsal zone of periventricular hypothalamus	TTBr	tractus tectobulbaris rectus
Hv	ventral zone of periventricular hypothalamus		
IMRF	intermediate reticular formation	VIII	octaval nerve
IRF	inferior reticular formation		

6 Functional anatomy of the zebrafish brain: a comparative evaluation

Most basic to an understanding of functional neuroanatomy are neuronal connections in the CNS. Here, only connections established with experimental neuronal tracing or with degeneration experiments in various teleost species (including goldfish and carp) are considered. In general, connections are meant to be ipsilateral unless otherwise mentioned.

Sensory Systems in the teleostean CNS

Olfaction
As in all vertebrates, the only primary sensory receptor cells in teleosts are located in the olfactory mucosa. The axons of these cells represent the primary olfactory projections (fila olfactoria, nervus olfactorius, I) and reach the glomerular layer (GL) of the olfactory bulbs (Nieuwenhuys and Meek, 1990). In the zebrafish, the development of the olfactory organ and of the primary olfactory projections from the placodal stage to the adult configuration has been documented with transmission and scanning electron microscopy (Hansen and Zeiske, 1993). Compared with several hundred glomeruli in larger teleost species, the adult zebrafish has a relatively low total number of 80 glomeruli per olfactory bulb. Of these at least 22 are individually identifiable glomeruli (Baier and Korsching, 1994). Olfactory receptor cells that provide input to a small area of the olfactory bulb (1% of the total volume of the glomerular layer) are distributed evenly all over the olfactory mucosa in the rainbow trout (Riddle and Oakley, 1991). Non-overlapping sets of such widely distributed receptor cells appear to project to different glomeruli in the trout. If these sets of receptor cells each have a limited number of different receptor types, a single olfactory glomerulus may be processing information derived from only a very few receptor types.
Another cranial nerve, the terminal nerve (nervus terminalis), runs together with the olfactory nerve. In most teleosts, the terminal nerve ganglion cells lie in or near the ventral olfactory bulb. These ganglion cells have a periph-eral dendrite that sometimes reaches into the olfactory mucosa and a central axon that always projects beyond the olfactory bulbs into the ventral telencephalon, preoptic diencephalon, or contralateral retina (Bartheld, 1987). The functional significance of the terminal nerve is not entirely clear, although there is good evidence that it may be involved in pheromone detection and in mediating sexual behaviors such as sperm release (Demski and Dulka, 1984; Demski and Sloan, 1985; Dulka et al., 1987). In the zebrafish, the terminal nerve and its ganglion cells have not yet been identified.

The teleostean secondary olfactory projections originate in the large mitral cells of the olfactory bulb and run in the lateral and medial olfactory tracts (LOT/MOT). While the MOT (which includes the terminal nerve fibers) appears to carry information related to sexual behaviors, the LOT mediates feeding behavior and alerting responses (for further reference to physiological and behavioral aspects of olfaction in fish, see: Hara, 1992). Secondary olfactory projections in teleosts are ipsilateral (with a small contralateral component) and reach most nuclei in the area ventralis telencephali, a caudal ventro-lateral part of the area dorsalis telencephali, as well as the preoptic and posterior tubercular regions in the diencephalon (Nieuwenhuys and Meek, 1990). In the goldfish (Bartheld et al., 1984; Levine and Dethier, 1985), the most dense secondary olfactory terminals in the ventral telencephalic area are located in the central (Vc), ventral (Vv), and dorsal (Vd) nuclei, while in the dorsal telencephalic area, the dorsal posterior zone (Dp) and nucleus taeniae (NT) are the major recipients of projections. Secondary olfactory terminals are also found in the anterior preoptic region and the posterior tuberal nucleus of the goldfish diencephalon. The olfactory bulb, in turn, receives projections from the secondary olfactory (and additional) centers in the telencephalon as well as from the contralateral olfactory bulb.

Vision
Most teleosts are highly visually guided animals, and some

of their capabilities involving this sensory modality are impressive (Douglas and Djamgoz, 1990), e.g. presence of four cone types (tetrachromacy) in goldfish color vision (Neumeyer, 1992), or maintenance of size and color constancy of objects (Douglas and Hawryshyn, 1990). While the retinotectal system is unquestionably predominant for teleostean vision, there is considerable variability in how the rest of the visual system, especially the pretectum, is organized in various teleost taxa. This variability has caused considerable difficulty in understanding the structure and function of the teleostean visual system and in interpreting its phylogeny (for reviews, see: Northcutt and Wullimann, 1988; Wullimann et al., 1991b).

In contrast to other sensory nerves and sensory organs, which are all embryonic derivatives of neural crest and placodes, the retina and optic nerve (ON, nervus opticus, II) are embryonically derived from the neural tube and thus constitute part of the CNS. As in other vertebrates, five major central nervous areas receive primary retinal input (mostly contralaterally) in teleosts: 1) the optic tectum, 2) the thalamus, 3) the pretectum, 4) the accessory optic system, and 5) the preoptic area (for references on primary visual projections in teleosts, see: Northcutt and Wullimann, 1988).

The ganglion cells of the retina project topographically onto the optic tectum (TeO) where they form several bands of terminals (Bartheld and Meyer, 1987). In the goldfish (Meek, 1983; Northcutt, 1983), the most peripheral retinal fibers are located in the superficial white and grey zone; where a thin band is located in layer 14, and a much thicker band in layers 8 to 12. A third and fourth band of retinal input are seen at the boundary zone between layers 5 and 6 within the central zone, and in layer 4 within the deep white zone. The teleostean optic tectum is not exclusively a visual structure, but it has been investigated mostly in that respect (see this chapter: Tectum opticum). There can be no question that the optic tectum is the major visual center in the teleostean brain. It is there that visual information concerning movement, shape, and color are analyzed (Guthrie, 1990).

However, distinct retinal terminal fields are also present lateral to the anterior (A), intermediate (I), and ventrolateral (VL) thalamic nuclei, but little is known concerning their physiology (Schellart, 1990). Similar to the situation in tetrapods, two higher-order visual pathways to the telencephalon arise in the teleostean dorsal thalamus, one via the anterior thalamic and a second via the dorsal posterior thalamic nucleus (DP; Echteler and Saidel, 1981). The anterior thalamic nucleus is a primary retinal target, and the dorsal posterior thalamic nucleus is in receipt of tectal input. The pathways ascending from these two nuclei to the telencephalon thus resemble the geniculate and extrageniculate visual systems of mammals and their homologues in other tetrapods. However, the telencephalic projections from these dorsal thalamic nuclei are weak in teleosts in comparison with those of the non-teleost actinopterygian fish *Polypterus* (Northcutt, 1981a), suggesting that these pathways may have become reduced in teleosts.

In some teleosts, a third ascending visual pathway to the telencephalon arises in the most anterior, tecto-recipient part of the preglomerular area («nucleus prethalamicus» of other authors, see: Northcutt and Wullimann, 1988). In cyprinids, most preglomerular nuclei project to the telencephalon, but there is no strong tectal input – typical of «nucleus prethalamicus» – to any of these nuclei (Grover and Sharma, 1979; Luiten 1981). Thus, cyprinids appear to lack a «nucleus prethalamicus». Because of its strong tectal input, the magnocellular superficial pretectal nucleus (PSm) of cyprinids – although it does not project to the telencephalon – has been misidentified as «nucleus prethalamicus».

The pretectum is the most variable visual subsystem in teleosts. Although major patterns of pretectal organization in different teleosts have been recognized, very little is known concerning the underlying functional context. The most consistently observed retinorecipient teleostean pretectal nucleus is the central pretectal nucleus (CPN). This nucleus is likely to be the homologue of the lentiform mesencephalic nucleus of other non-mammalian vertebrates and, thus, may be functionally closely associated with the accessory optic nuclei in oculomotor reflexes (see below). Also, little interspecific variation is seen in the retinorecipient dorsal periventricular pretectal nucleus (PPd). Its functional context is unknown. The superficial pretectum is highly variable in teleosts, both in the number of nuclei and in their degree of structural differentiation. An elaborate pattern of pretectal organization is present in the most derived group of teleosts, the percomorphs. There it involves a major visual pathway from the retinorecipient parvocellular superficial pretectal nucleus (PSp) via nucleus glomerulosus to the hypothalamic inferior lobes (Sakamoto and Ito, 1982; Striedter and Northcutt, 1989; Wullimann and Meyer, 1990; Wullimann et al., 1991b; for a brain atlas representative for such teleosts, see: Anken and Rahmann, 1994). In addition, this pathway is paralleled by a visual input via nucleus corticalis to nucleus glomerulosus. Comparative studies suggest that a similar pretectal visual circuitry may have existed at the outset of teleostean evolution (Wullimann et al., 1991b). The size and degree of differentiation of nuclei and tracts involved in the retino-pretecto-hypothalamic pathway is astonishing and represents a unique specialization of teleosts. Although electrophysiological evidence suggests a role of this visual subsystem in the detection of moving objects (Rowe and Beauchamp, 1982; Williams and Vanegas, 1982), better knowledge of its functional context is necessary before the visual system of teleosts can be fully evaluated.

Cyprinids have a distinctly altered visual circuitry compared with that just described (Northcutt and Wullimann, 1988). They have a rather small PSp and lack both nucleus corticalis and nucleus glomerulosus. Unfortunately, in some of the literature on the gustatory system, the (preglomerular) tertiary gustatory nucleus (TGN) has been misidentified as nucleus glomerulosus (see this chapter: Gustation). A large tectorecipient magnocellular superficial pretectal nucleus (PSm) projects to the mammillary body (CM) and nucleus lateralis valvulae (NLV) in cyprinids (Northcutt and Braford, 1984; Striedter and

Northcutt, 1989; Ito and Yoshimoto, 1990; Yoshimoto and Ito, 1993), while in percomorphs the same-named nucleus projects to nucleus isthmi (NI). Thus, it appears that in cyprinids the connections of the magnocellular superficial pretectal nucleus are entirely different from those of their percomorph counterparts, and that the visual circuits running via the parvocellular superficial pretectal nucleus are highly reduced.

The dorsal and ventral accessory optic nuclei (DAO/ VAO) of teleosts receive retinal input. They are involved in optokinetic oculomotor reflexes (see this chapter: Motor nuclei of cranial nerves, Cerebellum). The retinorecipient (preoptic) suprachiasmatic nucleus (SC), judged from its role in other vertebrates, may be involved in circadian rhythm generation and control. Additional retinal fibers terminate in the preoptic region lying dorsal and caudal to the the suprachiasmatic nucleus (Butler and Saidel, 1991; Medina et al., 1993; Northcutt and Butler, 1993). It is interesting to note that whereas the accessory optic nuclei are easily lost in many vertebrates with a secondarily reduced visual system (e.g. in blind cave fish), the suprachiasmatic nucleus is not.

In the zebrafish, the morphology of the primary visual nuclei corresponds to the cyprinid pattern of pretectal organization and is, thus, also likely to be secondarily reduced. It is notable that the ventral accessory optic nucleus is large in the zebrafish compared with the minute nucleus seen in goldfish. Although primary retinal terminal fields have recently been described in embryonic and larval zebrafish (Burrill and Easter, 1994), the development of primary visual areas into their adult configuration is still unknown.

Mechanoreception

The term mechanoreception is used in this text exclusively for mechanosensory signals sensed by hair cells in neuromasts of the lateral line system and not for the tactile component of the trigeminal somatosensory system. Mechanosensory information reaches the brain via the lateral line nerves (LL; nervi lineae lateralis). In most aquatic vertebrate groups, the lateral line nerves encode a second sensory modality – electroreception – which is absent in most teleosts including cyprinids and will therefore not be considered here. The mechanosensory neuromasts detect the relative movement (acceleration) between water and animal at low frequencies (1–200 Hz) and at relatively short distances (1–2 body lengths) in various biological contexts, such as prey localization, navigation, and schooling behavior (for reviews, see: Coombs et al., 1989).

Lateral line nerves are characterized by distinct embryonic origins (placodes) and by adult sensory ganglia, receptor organs (neuromasts), as well as central projection nuclei, which are all different from those of other cranial nerves (McCormick, 1982; Northcutt, 1989). They are, thus, cranial nerves in their own right and do not represent the special somatosensory component of the facial and vagal nerves. In teleosts, the anterior and posterior lateral line nerves (ALLN/PLLN) enter the brain as anterior and posterior roots – as seen in the zebrafish – and they project to a dorsal medullary area between cerebellum (Ce) and vagal lobe (LX). This mechanoreceptive area is divided into a medial and a caudal octavolateralis nucleus (MON, CON). Additional lateral line projections always reach the cerebellar granular eminence (EG) and, in a few species, corpus and valvula cerebelli (for literature on primary lateral line projections in teleosts, see: Wullimann et al., 1991a).

Second order projections from these primary nuclei ascend in the lateral longitudinal fascicle (LLF) and terminate bilaterally, albeit with a stronger contralateral component in the ventrolateral nucleus of the torus semicircularis (TSvl Knudsen, 1977; McCormick, 1989), which in turn projects to the lateral preglomerular nucleus (PGl; Echteler, 1984; Murakami et al., 1986b; McCormick, 1989; Striedter, 1991; McCormick and Hernandez, 1996). This thalamic nucleus provides one of the strongest and interspecifically most consistent inputs to the area dorsalis telencephali, though to slightly variable subregions in different species (Dm/Dd/Dl/Dc; Murakami et al., 1986b;

Striedter, 1992). For example, in catfish and in cyprinids, the lateral, medial, and central zones of the area dorsalis telencephali all receive input from the lateral preglomerular nucleus, but the dorsal zone only does so in catfish.

There are descending projections within the teleostean mechanosensory system. In most species investigated, the medial and central zones of the area dorsalis telencephali project back onto the lateral preglomerular nucleus (PGl). However, there are no descending projections from PGl to the torus semicircularis. The torus semicircularis, a structure involved in mechanosensory processing, projects via a brain stem nucleus, the preeminential nucleus, to the primary sensory medial octavolateralis nucleus (Finger, 1986; Striedter, 1991; McCormick and Hernandez, 1996). The preeminential nucleus also receives input from the medial octavolateralis nucleus.

The mechanosensory system has efferent neurons in the vicinity of the facial motor nucleus (see this chapter: Motor nuclei of cranial nerves).

In the zebrafish, the mechanosensory circuitry has not been investigated experimentally. However, most of the nuclei of the mechanosensory system, such as the medial and caudal octavolateralis nuclei, the ventrolateral nucleus of the torus semicircularis, and the lateral preglomerular nucleus, are apparent morphologically. Unambiguous identification of the preeminential nucleus is lacking in the zebrafish, although the (unlabeled) nucleus ventral to the trochlear nerve and dorsal to nucleus isthmi (cross section 213) is a good candidate.

Audition

The auditory capabilities of many, if not all, teleosts are impressive (for reviews see: Webster et al., 1992; Popper and Fay, 1993). Auditory signals are perceived for considerably greater distances than are mechanosensory signals, and the perceived frequency ranges up to 3000 Hz. Rather surprising is the fact that synthetic listening (several components of a complex signal grouped together) and analytic listening (identification of one component of a complex signal) occur in goldfish (Fay, 1992), since these behaviors were long thought to be unique to mammals. The peripheral auditory receptor organs are one or more otolithic end organs of the inner ear. Sensory fibers innervating hair cells of various end organs of the inner ear form the octaval nerve (nervus octavus, VIII), which encodes vestibular as well as auditory information. In contrast to the older view of a common primary sensory octavolateralis area receiving both lateral line and octaval nerve projections, it is now accepted that the primary projections of these nerves are segregated into a dorsal mechanosensory (lateral line) column (see above) and a ventral octaval column (McCormick, 1982; 1992). The octaval nerve projects to five nuclei comprising this octaval column, which are the anterior (AON), magnocellular (MaON), descending (DON), tangential (T), and posterior octaval (PON) nuclei. Similar to the lateral line nerves, the octaval nerve additionally project to the cerebellar granular eminence (for literature on primary octaval projections, see: Wullimann et al., 1991a), where the octaval input is spatially segregated from the lateral line input. The corpus and valvula cerebelli generally do not receive primary octaval input. A small zone of overlap between primary lateral line and octaval projections exists in a limited part of the teleostean magnocellular octaval nucleus. Moreover, in the herring and in the eel, octaval projections also terminate in part of the (mechanosensory) medial octavolateralis nucleus (Meredith, 1985; Meredith et al., 1987). The functional significance of these cases of limited mechanosensory-octaval overlap of primary projections is unclear.

Different end organs in the utriculus, sacculus, and lagena can be specialized for audition in various teleosts (Bleckmann et al., 1991; Highstein et al., 1992; Popper and Fay, 1993). Correlated with these peripheral auditory specializations is some variation in the primary sensory central nuclei related to audition. In cyprinids, the dorsal part of the anterior octaval nucleus as well as the dorsomedial part of the descending octaval nucleus (Echteler, 1984; 1985; McCormick and Braford, 1994) have been identified as primary auditory centers.

Also in cyprinids, secondary octaval projections from these two nuclei ascend in the lateral longitudinal fascicle (LLF) and terminate bilaterally in the central nucleus (TSc) of the torus semicircularis (with a stronger contralateral component) and in a secondary octaval population of neurons (with a stronger ipsilateral component) (Echteler, 1984; McCormick and Hernandez, 1996). Interestingly, the secondary octaval population of neurons (SO) projects to the central nucleus of the torus semicircularis and may be homologous to the superior olive of mammals. The central nucleus of the torus semicircularis projects to the central posterior thalamic nucleus (CP; Echteler, 1984). Although it has been demonstrated that the central and the medial zones of the area dorsalis telencephali process auditory information (Echteler, 1985), in cyprinids, direct projections from the central posterior thalamic nucleus to these telencephalic regions have not so far been found (Wullimann and Meyer, 1993). The latter may thus receive already secondary auditory input from within the telencephalon. However, the central posterior thalamic nucleus undoubtedly projects to the telencephalon in some teleost species (Wullimann and Northcutt, 1990; Striedter, 1991), but its main telencephalic target may be within the area ventralis (Striedter, 1991).

The descending projections in the cyprinid auditory system include a projection from the secondary octaval population back to the descending octaval nucleus, as well as a projection from the central posterior thalamic nucleus and from the central zone of the area dorsalis telencephalic to the auditory torus semicircularis (Echteler, 1984; McCormick and Hernandez, 1996).

Additional auditory pathways from the torus semicircularis via the diencephalon to the telencephalon exist in teleosts. A pathway via the ventromedial thalamic nucleus in *Sebastiscus* (Murakami et al., 1986a) and a second one via the tuberal hypothalamus in catfish (Striedter, 1991) have been described. These pathways appear to be specializations of the species mentioned.

Similar to the mechanosensory system, the auditory system has efferent neurons in the vicinity of the facial motor nucleus (see this chapter: Motor nuclei of cranial nerves).

In the zebrafish, the central anatomy of the primary octaval nuclei and the higher-order auditory system is very similar to that just discussed for other cyprinids. However, except for the descending spinal projections of vestibular nuclei (see below), no experimental neuroanatomical data concerning the octaval system exist for zebrafish.

Vestibular sense

The peripheral receptors that mediate the sense of balance are found in the inner ear semicircular canal and otolithic end organs. Some of these sensory organs have a dual function in teleosts, since they are also involved in hearing (Popper and Fay, 1993; see above). Vestibular signals reach the nuclei of the octaval column from the peripheral receptors by way of the eighth cranial nerve (nervus octavus, VIII). Of the five primary octaval nuclei described above, the magnocellular, tangential and posterior octaval nuclei (MaON, T, PON) are likely to be exclusively vestibular. However, parts of the anterior and descending octaval nuclei (AON, DON) also receive vestibular information. All of the primary vestibular areas project to the spinal cord, in contrast to the auditory areas of the AON and DON (see this chapter: Descending spinal projections). Projections of primary vestibular nuclei to motor nuclei of extraocular muscles are discussed below (see this chapter: Motor nuclei of cranial nerves). Vestibular magnocellular, tangential and posterior octaval nuclei appear not to have ascending projections.

In the zebrafish as in other cyprinids, five octaval nuclei are present, and descending spinal projections of vestibular nuclei have been described (see this chapter: Descending spinal projections). However, further information regarding the vestibular central anatomy in this species is lacking.

Gustation

The gustatory system of fish can be differentiated from the olfactory system based primarily on its peripheral and central anatomy. Functionally, the two systems are harder to distinguish. Unlike in land vertebrates, the chemical cues for both systems are carried in aqueous medium. The gustatory system of fish is also active at long distance, as is the olfactory system. Furthermore, there is overlap between the two systems with respect to chemicals perceived, i.e. amino acids (for reviews, see: Hara, 1992).

Three cranial nerves, the facial (nervus facialis, VII), glossopharyngeal (nervus glossopharyngeus, IX), and vagal (nervus vagus, X), contact the multicellular peripheral receptor organs (taste buds) that are specialized to perceive environmental gustatory cues. These three cranial nerves project to the medullary gustatory (special viscerosensory) column. In fish with a well-developed taste system, this column consists of separate primary gustatory centers for each nerve, i.e. facial (LVII), glossopharyngeal (LIX), and vagal (LX) lobes (Luiten, 1975; Morita et al., 1980; Kanwal and Caprio, 1987; Puzdrowski, 1987). Projections of the glossopharyngeal nerve also reach part of the vagal lobe. The vagal nerve further projects to the medial funicular nucleus (MFN) and the commissural nucleus of Cajal (NC; = general viscerosensory nucleus). While taste buds on the head and body trunk (extraoral system) are innervated by the facial nerve, those in the oropharynx and on the gill arches (intraoral system) are innervated by the glossopharyngeal and vagal nerves. Hypertrophy of the gustatory system has occurred in several groups of teleosts independently, the two bestinvestigated groups being the silurids (catfish) and the cyprinids (carp, goldfish). The former have a relatively better developed extraoral system, whereas the latter have emphasized the intraoral system. This difference is also reflected in the central nervous representation of the different gustatory components in the two teleost groups. The facial lobe is larger and more complex in silurids (Hayama and Caprio, 1989), and the vagal lobe is larger and more complex in cyprinids (Morita and Finger, 1985). Most silurids have extensive barbels densely populated with taste buds, and they use this extraoral system to search for food, whereas they use their intraoral taste system to selectively ingest food. Cyprinids rely much more on their intraoral gustatory system, especially the palatal organ located in the oropharyngeal roof, for sorting food (Sibbing and Uribe, 1985). The entire oropharyngeal cavity, including the palatal organ and gill arches, is topographically represented in the vagal lobe (viscerotopy), with sensory and motor components in radial (laminar) register (Morita and Finger, 1985), allowing for point-to-point reflex arches involving oropharynx and vagal lobes. Interactions between gustatory and tactile systems exist at several levels in cyprinids and silurids:

1. Primary sensory trigeminal and facial projections originating in the mandibular periphery overlap in a limited ventral area of the facial lobe (Kiyohara et al., 1986; Puzdrowski, 1987; 1988). Similarly, the medial funicular nucleus of cyprinids receives primary projections from the trigeminal as well as the vagal nerves (Morita et al., 1980; Puzdrowski, 1988).
2. The facial, but not the vagal, lobe projects to the medial funicular nucleus in silurids and cyprinids (Finger, 1978a; Morita et al., 1980).
3. Apart from these central interactions of the gustatory and somatosensory systems, the teleostean vagal and facial fibers themselves encode tactile stimuli (Kiyohara et al., 1985; Marui et al., 1988; Kanwal and Caprio, 1988).

A common pattern of ascending gustatory connections is shared by all teleosts and can be concluded to represent the ancestral pattern. The primary gustatory nuclei project via the secondary gustatory tract (SGT) to a secondary gustatory nucleus (SGN) in the isthmus, which in turn projects to the hypothalamic inferior lobes (IL) and to a tertiary gustatory nucleus (TGN) within the medial preglomerular area (Finger, 1978a; Morita and Finger, 1985; Morita et al., 1980; 1983; Wullimann, 1988; Lamb and Caprio, 1993). As Braford and Northcutt (1983) have already pointed out, this preglomerular tertiary gustatory nucleus has been misidentified as «nucleus glomerulo-

sus» in cyprinids. The latter term definitely should be restricted to the visually related nucleus seen in more derived teleosts (see this chapter: Vision). Both silurids and cyprinids elaborate on this basic gustatory circuitry in slightly different ways. Facial, glossopharyngeal, and vagal lobes, as well as the secondary gustatory nucleus, are greatly enlarged and histologically more complex in these fish compared with species that show the ancestral pattern. However, the emphasis is on the facial lobe in silurids and on the vagal lobe in cyprinids. While cyprinids have a single preglomerular tertiary gustatory nucleus, silurids show two such preglomerular centers, the nucleus of the lateral thalamus, which is homologous to the cyprinid TGN, and a more posteriorly located nucleus lobobulbaris (Morita et al., 1980; 1983; Lamb and Caprio, 1993). Different tertiary gustatory centers, the inferior lobe in cyprinids and the lobobulbar nucleus in silurids, develop extensive descending connections to the (primary gustatory) facial and vagal lobes (Morita et al., 1983; Morita and Finger, 1985). In cyprinids, descending connections also originate in the area of the posterior thalamic nucleus (P). Thus, although the latter nucleus does not receive tertiary gustatory input in cyprinids, it may be homologous to the lobobulbar nucleus of silurids. Furthermore, the nucleus of the lateral thalamus in silurids (but not the preglomerular tertiary gustatory nucleus in cyprinids) projects back to the secondary gustatory nucleus.

Gustatory information generally does not appear to ascend directly from the thalamic level to the telencephalon, since in cyprinids the preglomerular tertiary gustatory nucleus and the inferior lobe do not project to the telencephalon (Wullimann and Meyer, 1993). However, in silurids the (tertiary gustatory) nucleus lobobulbaris projects to the medial zone of the dorsal telencephalic area (Dm) and the central nucleus of the inferior lobe projects to the central zone of the dorsal telencephalic area (Dc; Kanwal et al., 1988; Lamb and Caprio, 1993). Thus, these telencephalic connections appear to be a derived condition for silurids.

In summary, different elaborations of ascending and descending connections in the gustatory system distinguish cyprinids and silurids from other teleosts. These differences are likely to be related to behavioral differences in food handling.

The central anatomy of the gustatory system of the zebrafish is similar to that seen in goldfish in terms of morphology of the primary, secondary, and tertiary gustatory centers. However, in terms of connectivity, nothing is known about the primary and higher-order gustatory system in the zebrafish. Several interesting neurobiological problems regarding the zebrafish gustatory system await to be resolved. The relative contribution of extraoral versus intraoral gustatory systems, the presence or absence of viscerotopy and corresponding motor layer in the vagal lobes, and the degree of hidden lamination in the vagal lobes require investigation.

General visceral sense

In addition to the special viscerosensory modality (gustation), the teleostean vagal nerve also encodes general viscerosensory stimuli from the viscera to the CNS, notably to the nucleus commissuralis of Cajal (NC). In the goldfish this nucleus has lateral and medial subdivisions, the lateral one receiving general viscerosensory input from the posterior pharynx (region of the chewing organ immediately rostral to the esophagus) and the medial one from the gastrointestinal tract (Morita and Finger, 1987). The vagal motor neurons (NXm) are also segregated into lateral and medial populations; the lateral one innervates the posterior pharynx, and the medial ones innervate the gastrointestinal tract. A set of even more medially lying motoneurons innervate the heart. Similar to the vagal lobe (see above), reflex arches exist for the posterior pharynx and the viscera via the related sensory and motor nuclei (Goehler and Finger, 1992). The motoneurons related to the heart and the medial visceromotor subnucleus subserving the gastrointestinal tract are the homologue of the mammalian (parasympathetic) nucleus dorsalis nervi vagi. The lateral subnucleus innervating the posterior pharynx plus the motor layer in the vagal lobe innervating the oropharynx and gill arches are together homologous to the mammalian nucleus ambiguus. The nucleus commissuralis of Cajal together with the sensory layers of the vagal lobe are homologous to the mammalian nucleus solitarius.

Secondary general viscerosensory projections in silurids and cyprinids ascend in parallel with the secondary gustatory projections and terminate in a discrete area ventrally adjacent to the secondary gustatory nucleus (Finger and Kanwal, 1992). Tertiary general visceral projections are not known.

The zebrafish (general viscerosensory) nucleus commissuralis of Cajal and vagal motor nucleus appear neuroanatomically similar to those in the goldfish. However, their detailed relationship to the periphery must be established by experimental work, since morphological subdivisions are not overtly apparent. This is of particular importance with regard to the vagal motor nucleus, since it contains not only the parasympathetic motor neurons and the pharyngeal general visceromotor neurons, but possibly the special visceromotor neurons innervating the branchial arches as well.

Somatosensory system

Little is known about the ascending somatosensory system in teleosts. In the salmon *Oncorhynchus*, spinal axons ascend in the lateral funiculus and reach medullary cranial nerve motor nuclei, vagal lobe, reticular formation, cerebellum, and torus semicircularis. Ascending spinal fibers in the dorsal funiculus reach only to the caudal medulla oblongata (Oka et al., 1986b). This resembles the anterolateral and dorsal funicular ascending spinal systems, respectively, in mammals. In the scorpaenid *Sebastiscus*, ascending projections from the dorsal horn at the level of the obex reach the thalamus (VM/VL/DP) in addition to most targets reported for the salmon (Murakami and Ito, 1985; Ito et al., 1986). In the sea robin *Prionotus*, ascending projections reach from the caudal medulla to the (thalamic) preglomerular area (Finger, 1981). Thus, teleosts appear to have a relay center at the spinal cord-brain stem junction for ascending somatosensory fibers similar to the dorsal column cuneate and gracile nuclei in mammals, in addition to ascending spinal projections that are not relayed at this level.

In the zebrafish, ascending spinal projections originating caudal to the level of the first dorsal root appear similar to those described in salmon, and they also reach the ventral and dorsal thalamus (T. Becker, M. F. Wullimann, R. R. Bernhardt, C. G. Becker, M. Schachner; unpublished observations). This indicates that at least some ascending spinal projections reach the thalamus without being relayed at the spinal cord-brain stem junction. Recently, spinothalamic projections have also been demonstrated in amphibians (Munoz et al., 1994). Thus, spinothalamic projections may be present in anamniotes as well as in amniotes.

Complementary to the ascending somatosensory system that subserves the trunk periphery, the sensory component of the trigeminal nerve (nervus trigeminus, V) is concerned with somatosensation in the head. These sensory trigeminal projections have been investigated in the carp and the goldfish (Luiten, 1975; Puzdrowski, 1988). After entering the rostral medulla oblongata, the sensory root of the trigeminal nerve (Vs) divides into a rostral bundle and the descending trigeminal root (DV). The former terminates in the isthmic primary sensory trigeminal nucleus (NVs) and the DV descends in the medulla oblongata providing somatosensory input to the nucleus of the descending trigeminal root (NDV) and, further caudally, to the medial funicular nucleus (MFN). The (sensory) mesencephalic nucleus of the trigeminal nerve will be discussed in the next section.

In terms of higher-order connections of the somatosensory system, it is noteworthy that in *Sebastiscus*, the thalamic region indicated above (VM/VL/DP) receives converging input from the isthmic primary sensory trigeminal nucleus, the nucleus of the descending trigeminal root, and from spinal ascending somatosensory fibers, and this thalamic region, in turn, projects to the dorsal telencephalic

area (Dc/Dd/Dm; Ito et al., 1986). Thus, the entire somatosensory body surface appears to be represented in a limited part of the thalamus, and this information is relayed from here farther on to the telencephalon.

Motor and premotor systems in the teleostean CNS

Motor nuclei of cranial nerves

In cyprinids, the motoneurons of the oculomotor nerve (NIII) innervate four of the six extraocular eye muscles (rectus superior, inferior and internus (= medialis), obliquus inferior), the trochlear motor nucleus (NIV) innervates the obliquus superior, and the abducens motor nucleus innervates the rectus externus (= lateralis) (Luiten and Dijkstra - de Vlieger, 1978; Graf and McGurk, 1985). In the zebrafish and other cyprinids, the abducens motor nucleus consists of small rostral and caudal subnuclei (NVIr/NVIc). In the goldfish, the caudal subnucleus receives a reticular formation input that may be related to fast eye movements, in contrast to the slow eye movements effected by the rostral subnucleus (Allum et al., 1981). The oculomotor nucleus also receives input from reticular formation neurons (Torres et al., 1992), and the general functional context of these reticular projections is eye-body motor coordination. As part of the vestibuloocular reflex circuitry, primary octaval nuclei (ipsilateral AON, contralateral T and DON) project to abducens and oculomotor nuclei (Allum et al., 1981; Torres et al., 1992). Neurons in both parts of the contralateral abducens motor nucleus project to the oculomotor nucleus and are likely to be involved in horizontal conjugate eye movements performed simultaneously by the lateral rectus of one eye and the medial rectus of the other eye. In cyprinids, connections from the primary visual dorsal accessory optic nucleus to the oculomotor nucleus are absent (Torres et al., 1992). However, such projections are present in more visually guided teleosts (Finger and Karten, 1978; Uchiyama et al., 1988) and presumably are involved in the optokinetic ocular reflex.

The trigeminal and facial motor nuclei are involved in a variety of behaviors, from feeding and respiration to aggression, sexuality, and brood care. Here, the motor control of the rhythmic generation of water flow during respiration is considered in order to exemplify the neuronal circuitry of these motor nuclei. In cyprinids, the rostral portions of the trigeminal and facial motor nuclei innervate those muscles of the mandibular and hyomandibular arch that are active during the contraction of the buccal and opercular cavities, while the caudal portions of those motor nuclei innervate the antagonistic muscles performing the expansion of the respiratory cavities (Luiten, 1976; Song and Boord, 1993). Trigeminal and facial motor nuclei are functionally closely linked ipsi- and contralaterally with regard to respiratory coordination, since they extend dendrites reciprocally and bilaterally towards each other. Also, these motor nuclei receive terminals bilaterally from respiratory-active reticular formation neurons (Luiten and van der Pers, 1977). Further, both the trigeminal and the facial motor nuclei receive input from primary sensory trigeminal nuclei (NVs, NDV) and from neurons in the ventral facial lobe allowing for integrated reflexes towards tactile and gustatory cues. In mammals, the mesencephalic nucleus of the trigeminal nerve (MNV) contains proprioceptive neurons for the masticatory muscles, i.e. the central fibers of these neurons synapse on motor trigeminal neurons (monosynaptic masticatory reflex). This is not so in cyprinids, where proprioceptive neurons for head muscles involved in feeding and respiration (and in oculomotor movements) are located in the peripheral trigeminal and facial nerve ganglia (Luiten, 1979). Instead of being proprioceptive, the sensory neurons of the cyprinid MNV appear to be concerned with perioral (exteroceptive) information.

In the zebrafish, dorsal and ventral (NVmd/NVmv), rather than rostral and caudal, motor trigeminal subnuclei are present, and thus only the caudal part of the ventral subnucleus may be involved in respiratory expansion. The facial motor nucleus (NVIIm) of the zebrafish is morphologically not separated into rostral and caudal divisions.

Detailed tracing experiments are necessary to establish the functional subdivisions in the zebrafish described above for other cyprinids.

While the lateral line mechanosensory and octaval systems do not have motor nuclei that innervate muscles, they do have efferent neurons that innervate their peripheral sense organs. All teleosts have one efferent nucleus that lies in the midline of the rhombencephalon at the level of the facial motor nucleus and innervates both mechanosensory and inner ear end organs (Roberts and Meredith, 1989; 1992). The goldfish has two efferent neuronal populations in that region and a third one in the diencephalic periventricular area of the posterior tuberculum (Zottoli and van Horne, 1983; Puzdrowski, 1989). These three nuclei have also been documented in the larval zebrafish (Metcalfe et al., 1985), but the nuclei are not distinct in the adult zebrafish brain and their identification thus depends on tracing experiments.

Because of their intricate anatomical and functional relationship with the gustatory and general viscerosensory system, the motor nuclei of the glossopharyngeal and vagal nerves have already been treated there (see this chapter: Gustation and General visceral sense).

Descending spinal projections

Descending spinal projections in the salmon and in the goldfish (Oka et al., 1986a; Prasada Rao et al., 1987) course caudally within the bulbospinal tract (TBS), the medial longitudinal fascicle (MLF), and the vestibulo-spinal tract (TVS). These projections originate in the Mauthner cell, in all three parts of the reticular formation, in the superior raphe (SR), in the nucleus of the medial longitudinal fascicle (NMLF) and nucleus ruber (NR), in the preoptic region, and in some of the octaval nuclei. Interestingly, only octaval nuclei related to vestibular input (MaON, T, AON, ventrolateral part of DON; McCormick and Braford, 1994) descend to the spinal cord. The acoustically related dorsomedial part of DON does not. In the zebrafish, similar descending spinal projections have recently been described (Becker et al., 1995), and a detailed analysis of reticulospinal neurons exists for the embryonic (Mendelson, 1986), larval (Metcalfe et al., 1986), and adult (Lee and Eaton, 1991) zebrafish. Furthermore, the development of spinal motor neurons has been studied in the zebrafish (Myers, 1985; Eisen et al., 1986).

Reticular formation

The reticular formation constitutes a most complexly wired neuronal network extending throughout the medulla oblongata and into the tegmentum (see chapter 4: Medulla oblongata). The reticular formation has reciprocal connections with the optic tectum (Grover and Sharma, 1979; 1981; Luiten, 1981; Bosch and Paul, 1993) as well as with the cerebellum (Finger, 1983; Wullimann and Northcutt, 1988); there are reticular formation neurons that project to the spinal cord (Prasada Rao et al., 1987; Lee et al., 1993a), and reticular formation neurons in the midline column (serotoninergic superior raphe) project to the telencephalon (Murakami et al., 1983; Wullimann and Meyer, 1993). Some additional medullary structures may be considered part of the reticular formation, such as the Mauthner cell (Lee et al., 1993b) or the locus coeruleus (LC). As in all vertebrates, the LC includes noradrenergic neurons projecting to most brain areas, including the telencephalon (Ma, 1994a; 1994b). Some distinct tegmental nuclei are closely associated with (and may be considered part of) the reticular formation, such as the cerebellar-projecting perilemniscal nucleus (PL) and the spinal-projecting nucleus of the lateral lemniscus (NLL), as well as the red nucleus (NR), which projects to the spinal cord (Prasada Rao et al., 1987) and receives contralateral cerebellar input (Wullimann and Northcutt, 1988) in cyprinids.

The neuronal network just outlined constitutes the structural basis for the roles of the reticular formation in premotor functions as well as in functions of the ascending (noradrenergic, serotoninergic) activation systems (see this chapter: Telencephalon).

The development of a particular class of reticular formation neurons has been studied in the zebrafish (Kimmel et al., 1985).

Integrative centers in the teleostean CNS

Cerebellum

The teleostean cerebellum (Ce) is tripartite (see chapter 4: Cerebellum). The vestibulolateralis lobe comprising eminentia granularis (EG) and caudal lobe (LCa) is likely to be homologous to the vestibulocerebellum present in all vertebrates, since it receives primary octaval (presumably vestibular) as well as lateral line projections (see this chapter: Audition, Mechanoreception). The inputs to the teleostean corpus cerebelli (CCe) also are largely comparable to those of other vertebrates (Finger, 1978b; Wullimann and Northcutt, 1988). There is a climbing fiber input from a single source, the inferior olive (IO), as well as various additional, mossy fiber-like inputs originating in the spinal cord, in sensory medullary nuclei (MON/DON/NDV), in premotor centers (reticular formation, lateral reticular nucleus), and in the locus coeruleus (LC). Visual input to the corpus cerebelli comes from the pretectum (PPd/CPN), nucleus isthmi (NI), and accessory optic system (DAO/VAO). However, inputs to the corpus cerebelli from nucleus lateralis valvulae (NLV), dorsal tegmental nucleus (DTN), and telencephalorecipient nucleus paracommissuralis (PCN) appear to be special for teleosts. While the valvula cerebelli also receives input from the inferior olive and the locus coeruleus, the remaining input predominantly comes from the nucleus lateralis valvulae, the dorsal tegmental nucleus, and, only in goldfish, from the primary sensory trigeminal nucleus (NVs), eminentia granularis, and the preeminential nucleus (Wullimann and Northcutt, 1989). These differences in input to corpus and valvula cerebelli support the notion that the latter represents an evolutionary new part of the cerebellum (Nieuwenhuys, 1967).

The efferent cerebellar connections in teleosts arise from eurydendroid cells located in the same ganglionic layer as the Purkinje cells from which eurydendroid cells, in turn, receive input (Finger, 1983; Wullimann and Northcutt, 1988; Meek, 1992). Thus, unlike in cartilaginous fish and tetrapods, there are no deep (efferent) cerebellar nuclei in teleosts. The predominantly contralateral output of the teleostean cerebellum formed by the eurydendroid cells reaches, among other targets, motor and premotor centers, such as the oculomotor nucleus (NIII), nucleus ruber (NR), nucleus of the medial longitudinal fascicle (NMLF), and reticular formation.

The cytoarchitectonic properties of the teleostean cerebellar cortex and its input-output characteristics are so similar to other vertebrates that it probably subserves functions in motor learning and coordination also. A well-investigated example is the role of the lateral division of the valvula cerebelli (Val) in the dorsal light response in the goldfish (Mori, 1993; Yanagihara et al., 1993a; 1993b). However, the size of the valvula and representation of sensory systems in at least some teleost species (Lee and Bullock, 1984; Meek, 1992) raises the question whether additional cerebellar functions in teleosts should be anticipated. Interestingly, a discussion on possible new roles of the mammalian cerebellum (e.g. in cognition) has emerged recently (Leiner et al., 1991).

Tectum opticum

The teleostean optic tectum (TeO) is a cortex displaying laminae of neurons and neuropil that receive multimodal input from various sources (Meek, 1983; Northcutt, 1983). These include the retina (see this chapter: Vision) as well as additional visual centers (reviewed in Northcutt and Wullimann, 1988; Meek, 1990), such as the thalamus (anterior thalamic and ventrolateral/medial thalamic nuclei), pretectum (central pretectal and dorsal periventricular pretectal nuclei), and the nucleus isthmi, but also non-visual sources, such as the torus semicircularis or the telencephalon (central zone of area dorsalis telencephali). The ventrolateral nucleus of the torus semicircularis probably is the source of lateral line information to the optic tectum. The input from the nucleus isthmi to the tectum is part of a positive feedback system, likely in the context of alerting the tectum to significant visual events (King and Schmidt, 1991; 1993; Northmore, 1991). All inputs are radially segregated into various tangential bands within one or more tectal laminae. Many tectal neurons have a long dendritic arbor radially spanning the whole extent of the tectum. Some of these neurons may perform feature analysis of information gained from various visual and non-visual sources; the output of these neurons is to other visually related nuclei. Other neurons may integrate different multimodal information aspects; their output is to the premotor (reticular formation) and oculomotor nuclei (Niida et al., 1989; Meek, 1990). The origin of motor output is spatially patterned within the tectum, where largely non-overlapping areas (in the rostrocaudal and mediolateral planes) are responsible for ipsilateral turning, contralateral turning, or rolling of the eyes and the body (Guthrie, 1990). The cytoarchitectonic organization of the optic tectum, its segregated multimodal input, and the topographical representation of this input provide a neuronal machinery that is exquisitely designed for integrative orientation tasks such as object identification and location, and coordinated motor control.

Telencephalon

The smell-brain theory on vertebrate telencephalic evolution proposes that olfactory input dominates most, if not all, telencephalic areas in fish and that additional sensory systems invade the telencephalon only in tetrapods. This has turned out to be wrong for cartilaginous fishes (Ebbesson and Schroeder, 1971) or for teleosts, which have limited areas of secondary olfactory input in the telencephalon (see this chapter: Olfaction). Indeed, comparative research during the last 20 years has shown that there are more similarities in the organization of the teleostean and tetrapod telencephalon than previously believed (Nieuwenhuys, 1963; Vanegas and Ebbesson, 1976; Northcutt and Braford, 1980; Northcutt, 1981b; Murakami et al., 1983; Nieuwenhuys and Meek, 1990):

1. Most sensory systems reach the telencephalon in teleosts (see above), and there are strong reciprocal interconnections between thalamic and telencephalic sensory zones.

2. There are ascending activation systems to the teleostean telencephalon, such as the noradrenergic locus coeruleus (Ekström et al., 1986), the serotoninergic raphe nuclei (Kah and Chambolle, 1983), and possibly a dopaminergic substantia nigra-like system (see below), in addition to other inputs arising in the reticular formation.
3. In teleosts, two major telencephalic divisions (area dorsalis and ventralis telencephali), exhibiting characteristics reminiscent of pallial versus subpallial cell masses of tetrapods (histology, degree of differentiation, sensory input, neurotransmitter distribution), exist.

It is amazing that the teleostean telencephalon has these features, once believed to be characteristic only of the more highly evolved telencephalon, the ultimate integrative center, of tetrapods, especially of mammals. However, substantial questions remain regarding the recognition of homology between specific telencephalic areas, e.g. with regard to the dorsal pallium, hippocampus, or septum homologue in teleosts (Northcutt and Braford, 1980; Nieuwenhuys and Meek, 1990). Despite the topological distortion caused by eversion (see chapter 4: Telencephalon), connectional and immunohistochemical data have led to the recognition of likely candidates for the teleostean lateral pallium (olfactory cortex) and striatum. Since the posterior zone of the area dorsalis telencephali (Dp) is the strongest pallial recipient of secondary olfactory projections in all teleosts examined, it probably represents the lateral pallium of other vertebrates. In *Polypterus*, a non-teleost ray-finned fish, the dorsal nucleus of the area ventralis (Vd) displays Substance P-containing neurons that project to catecholaminergic (probably dopaminergic) ventral mesencephalic/posterior tubercular cells that in turn project back to Vd (Reiner and Northcutt, 1992). Also in teleosts, dopaminergic cells have been identified near or in the posterior tuberal nucleus (PTN; Roberts et al., 1989; Ekström et al., 1990). This indicates the presence in teleosts of a system similar to the striato-nigral system of tetrapods.

There are fundamental differences in telencephalic organization between teleosts and other vertebrates. A first is the developmental eversion of the telencephalic hemispheres (see chapter 4: Telencephalon), which applies uniquely to ray-finned fish. Its significance has yet to be determined. A second is the predominance of the posterior tuberculum versus the dorsal thalamus in relaying sensory information to the telencephalon in teleosts versus tetrapods. Teleosts may share this difference in comparison with tetrapods, with cartilaginous fish. Clearly, we are only beginning to fully understand the similarities and differences in telencephalic organization between teleosts and other major vertebrate groups.

7 Index of Latin Terms

(Section numbers refer to cross, **sagittal**, and *horizontal* sections)

Latin	English	Abbr.	Sect. No.	Text (page)
area dorsalis telencephali	dorsal telencephalic area	D	50, 60, 71, 85, 92, 98. 107, 114, 121	7, 8, 9, 89, 92, 97, 101
area dorsalis telencephali, zona centralis	central zone of dorsal telencephalic area	Dc	71, 85, 92, 98, 107; **15, 22, 34, 46, 53**; *1*	8, 92, 93, 95, 97, 100
area dorsalis telencephali, zona dorsalis	dorsal zone of dorsal telencephalic area	Dd	85, 92, 98; **22, 53**	8, 92, 97
area dorsalis telencephali, zona lateralis	lateral zone of dorsal telencephalic area	Dl	71, 85, 92, 98, 107, 114; **1, 9, 15, 22**; *1*	8, 92
area dorsalis telencephali, zona medialis	medial zone of dorsal telencephalic area	Dm	71, 85, 92, 98, 107. 114; **34, 46, 53**; *1*	8, 92, 93, 95, 97
area dorsalis telencephali, zona posterior	posterior zone of dorsal telencephalic area	Dp	71, 85, 92, 98, 107. 114; **1**; *9*, **15, 22, 34, 46,53**; *1, 18*	8, 9, 89, 101
area praeglomerulosa	preglomerular area	PG		11, 90, 95, 96
area ventralis telencephali	ventral telencephalic area	V	50, 60, 71, 85, 92, 98, 107; **15, 22, 34, 46, 53**; *1, 18, 26, 31*	7, 8, 9, 89, 93, 101
area ventralis telencephali, nucleus centralis	central nucleus of ventral telencephalic area	Vc	60, 71; **53**	8, 89
area ventralis telencephali, nucleus dorsalis	dorsal nucleus of ventral telencephalic area	Vd	50, 60, 71, 85; **34, 46**; *1, 18*	8, 89, 101
area ventralis telencephali, nucleus lateralis	lateral nucleus of ventral telencephalic area	Vl	60, 71; **15, 22, 53**; *18, 26, 31*	8
area ventralis telencephali, nucleus postcommissuralis	postcommissural nucleus of ventral telencephalic area	Vp	98, 107; **34, 46**; *1*	8
area ventralis telencephali, nucleus supracommissuralis	supracommissural nucleus of ventral telencephalic area	Vs	92; **34, 46, 53**; *1*	8, 9
area ventralis telencephali, nucleus ventralis	ventral nucleus of ventral telencephalic area	Vv	60, 71, 85; **34, 46, 53**; *18, 26, 31*	8, 9, 89
bulbus olfactorius	olfactory bulb	OB		7, 8, 89
canalis centralis	central canal	C	319, 363; **53**	15
cellula Mauthneri	Mauthner cell	MAC	230; *58*	15, 98, 99
cerebellum	cerebellum	Ce	31	7, 13, 14, 15, 16, 91, 92, 96, 99, 100

Latin	English	Abbr.	Sect. No.	Text (page)
chiasma opticum	optic chiasm	CO	107, 114; **53**; *58*	11
commissura ansulata	ansulate commissure	Cans	179; **34, 46, 53**; *58*	2, 16
commissura anterior, pars dorsalis	anterior commissure, dorsal part	Cantd	85, 92; **34, 46, 53**	9
commissura anterior, pars ventralis	anterior commissure, ventral part	Cantv	92; **34, 46, 53**; *18*	9
commissura cerebelli	cerebellar commissure	Ccer	213, 219; **9, 15, 22, 34, 46, 53**;*18*	13
commissura habenularum	habenular commissure	Chab	136; **46, 53**	1, 11
commissura horizontalis	horizontal commissure	Chor	121, 125, 127, 131, 136, 149, 153, 158, 162; **15, 22, 34, 46, 53**; *31, 46, 58*	11
commissura infima Halleri	commissura infima of Haller	Cinf	319; **46, 53**	15, 17
commissura intertectalis = commissura tecti	intertectal commissure = tectal commissure	Ctec	149, 153, 158, 162, 168, 173, 179, 185, 196, 201; **34, 46, 53**	1, 12
commissura nucleorum gustatorium secundariorum	commissure of the secondary gustatory nuclei	Cgus	213; **34, 46, 53**; *26*	17
commissura posterior	posterior commissure	Cpost	149, 153, 158, 162, 168; **22, 34, 46, 53**; *1, 18*	10, 11
commissura postoptica = commissura supraoptica	postoptic commissure = supraoptic commissure	Cpop	121, 125, 127; **34, 46, 53**; *58*	3, 11
commissura supraoptica = commissura postoptica	supraoptic commissure = postoptic commissure	Cpop	121, 125, 127; **34 46, 53**; *58*	3, 11
commissura tecti = commissura intertectalis	tectal commissure = intertectal commissure	Ctec	149, 153, 158, 162, 168, 173, 179, 185, 196, 201; **34, 46, 53**	1, 12
commissura tuberculi posterioris	commissure of the posterior tuberculum	Ctub	173; **46, 53**	1, 11
commissura ventralis rhombencephali	ventral rhombencephalic commissure	Cven	201, 204, 208, 213, 219, 223, 230, 237, 239, 251, 272; **34, 46, 53**, *46, 58, 76*	14, 15, 16
cornu dorsale	dorsal horn	DH	363; **53**	15, 96
cornu ventrale	ventral horn	VH	363	15
corpus cerebelli	cerebellar corpus	CCe	204, 208, 213, 219, 223, 230, 237, 239, 251; **9, 15, 22, 34, 46, 53**; *1, 18, 26*	13, 16, 92, 93, 99

Latin	English	Abbr.	Sect. No.	Text (page)
corpus mamillare	mammillary body	CM	173, 179, 185; **34, 46, 53**; *58, 76*	10, 11, 91
corpus pineale = epiphysis	pineal organ = epiphysis	E	136; **46, 53**	6, 9
crista cerebellaris	cerebellar crest	CC	237, 239, 251, 260, 272, 279; **15, 22, 34, 46, 53**; *18, 26, 31*	14
decussatio trochlearis	trochlear decussation	DIV	204; **34, 46, 53**	12
eminentia granularis	granular eminence	EG	219, 223, 230, 237, 239; **1, 9, 15**; *1, 18, 26, 31*	13, 14, 16, 92, 93, 99
epiphysis = corpus pineale	epiphysis = pineal organ	E	136; **46, 53**	6, 9
fasciculus lateralis telencephali	lateral forebrain bundle	LFB	92, 98, 107, 114, 121, 125, 127, 131, 136, 149; **15, 22, 34**; *1, 18, 26*	8, 9, 11
fasciculus longitudinalis lateralis = lemniscus lateralis	lateral longitudinal fascicle = lateral lemniscus	LLF	168, 173, 179, 185, 196, 201, 204, 208, 213, 219, 223; **9, 15, 22, 34**; *18, 26, 31, 46, 58*	13, 16, 92, 93
fasciculus longitudinalis medialis	medial longitudinal fascicle	MLF	173, 179, 185, 196, 201, 204, 208, 213, 219, 223, 230, 237, 239, 251, 260, 272, 279, 283, 290, 303, 319, 363; **46**; **53**; *26, 31, 46, 58, 76*	2, 15, 16, 98
fasciculus medialis telencephali	medial forebrain bundle	MFB	98, 107, 114, 121, 125, 127, 131, 136, 149; **34, 53**; *18*	9, 11
fasciculus retroflexus = tractus habenulo-interpeduncularis	habenulo-interpeduncular tract	FR	127, 131, 136, 149, 153, 158, 162, 168, 173, 179; **34, 53**; *1, 18, 26, 31, 46*	11
fibra Mauthneri	Mauthner axon	MA	237, 239, 251, 260, 272, 279, 283, 290, 303, 319, 363; **46**	15, 16
fibrae arcuatae internae	inner arcuate fibers	IAF	237, 239, 279; *58*	14, 16
formatio reticularis, pars inferior	inferior reticular formation	IRF	272, 279, 283, 290, 303, 319; **34, 46**; *76*	15, 16, 17, 98
formatio reticularis, pars intermedia	intermediate reticular formation	IMRF	230, 237, 239, 251, 260; **34, 46, 53**; *76*	14, 15, 17, 98
formatio reticularis, pars superior	superior reticular formation	SRF	196, 201, 204, 208, 213, 219, 223; **34, 46, 53**; *46, 58, 76*	12, 14, 15, 16, 98

Latin	English	Abbr.	Sect. No.	Text (page)
funiculus dorsalis	dorsal funiculus	Fd	363	15, 16, 96
funiculus lateralis, pars dorsalis	dorsal part of lateral funiculus	Fld	363; *58*	15, 16, 17
funiculus lateralis, pars ventralis	ventral part of lateral funiculus	Flv	303, 319, 363	15, 16, 17
funiculus ventralis	ventral funiculus	Fv	303, 319, 363	16
glandula pituitaria = hypophysis	pituitary = hypophysis	Pit	153, 158, 162, 168, 173	6, 10
griseum centrale	central gray	GC	196, 201, 204, 208, 213, 219, 223; **34, 46, 53**; *26, 31, 46*	12, 15
habenula	habenula	Ha	**34, 46, 53**; *1*	9, 11
hypophysis = glandula pituitaria	hypophysis = pituitary	Pit	153, 158, 162, 168, 173	6, 10
hypothalamus, pars tuberalis	tuberal hypothalamus	TH		10, 93
lemniscus lateralis = fasciculus longitudinalis lateralis	lateral lemniscus = lateral longitudinal fascicle	LLF	168, 173, 179, 185, 196, 201, 204, 208, 213, 219, 223; **9, 15, 22, 34**; *18, 26, 31, 46, 58*	13, 16, 92, 93
lacuna vasculosa areae postremae	vascular lacuna of area postrema	Vas	168, 173, 179; **46**	
lobus caudalis cerebelli	caudal lobe of cerebellum	LCa	230, 237, 239, 251, 260; **22, 34, 46, 53**; *1, 18, 26*	13, 99
lobus facialis	facial lobe	LVII	260, 272, 279, 283; **46, 53**; *46*	7, 14, 15, 94, 95, 98
lobus glossopharyngei	glossopharyngeal lobe	LIX	272	14, 94, 95
lobus inferior hypothalami	inferior lobe of hypothalamus	IL		10, 12, 91, 95
lobus vagi	vagal lobe	LX	279, 283, 290, 303; **22, 34, 46, 53**; *26, 31, 46*	7, 14, 15, 92, 94, 95 *96*
locus coeruleus	locus coeruleus	LC	204	15, 99, 101
medulla oblongata	medulla oblongata	MO		12, 13, 14, 16, 17, 96 97, 99
medulla spinalis	spinal cord	MS		1, 3, 4, 7, 12, 15, 16, 17, 94, 96, 97, 98, 99
nervi lineae lateralis	lateral line nerves	LL		13, 14, 91, 92, 93
nervi lineae lateralis anterioris	anterior lateral line nerves	ALLN	219, 223, 230, 237, 239; **1, 9, 15**; *46, 58*	13, 14, 92
nervus abducens	abducens nerve	VI		13, 14

Latin	English	Abbr.	Sect. No.	Text (page)
nervus facialis	facial nerve	VII	219, 223; **1**; *58*	13, 92, 94
nervus glossopharyngeus	glossopharyngeal nerve	IX	272	13, 14, 94
nervus lineae lateralis posterioris	posterior lateral line nerve	PLLN	251, 260, 272; **15**; *46, 58*	13, 92
nervus octavus	octaval nerve	VIII	230, 237, 239, 251, 260, 272; **9, 15**; *58, 76*	13, 14, 93, 94
nervus oculomotorius	oculomotor nerve	III	179; **1, 9, 15, 22, 34, 46, 53**; *46, 58*	12
nervus olfactorius	olfactory nerve	I		7, 89
nervus opticus (=II)	optic nerve	ON	107; **15, 22, 34, 46, 53**; *58*	11, 90
nervus trigeminus	trigeminal nerve	V	213; **1**; *58*	13, 14, 95, 97
nervus trochlearis	trochlear nerve	IV	201, 208, 213; **1, 9, 15, 22**; *26, 31*	12, 92
nervus vagus	vagal nerve	X	279, 283, 290; *46, 58*	13, 14, 92, 94, 95, 96
nucleus anterior thalami	anterior thalamic nucleus	A	131, 136; **34, 53**; *18*	9, 90, 100
nucleus centralis lobi inferioris hypothalami	central nucleus of the inferior lobe	CIL	173, 179, 185; **15, 22**; *76*	10, 95
nucleus centralis posterior thalami	central posterior thalamic nucleus	CP	149, 153; **34, 35**; *18, 26, 31*	9, 93
nucleus centralis tori semicircularis	central nucleus of semicircular torus	TSc	168, 173, 179, 185, 196, 201, 204, 208, 213; **9**; *1, 18*	12, 93
nucleus commissuralis Cajal	commissural nucleus of Cajal	NC	319; **46, 53**; *46, 58*	14, 17, 94, 96
nucleus descendens nervi trigemini	nucleus of the descending trigeminal root	NDV	239, 251, 260	13, 97, 98, 99
nucleus diffusus lobi inferioris hypothalami	diffuse nucleus of the inferior lobe	DIL	149, 153, 158, 162, 168, 173, 179, 185, 196, 201, 204; **9, 15, 22, 34, 46, 53**; *76*	10
nucleus dorsalis posterior thalami	dorsal posterior thalamic nucleus	DP	149, 153; **34, 53**; *18, 26*	9, 90, 96, 97
nucleus Edinger-Westphal	Edinger-Westphal nucleus	EW	173	12
nucleus entopeduncularis, pars dorsalis	entopeduncular nucleus, dorsal part	ENd	85, 92, 98; **22**; *18*	8
nucleus entopeduncularis, pars ventralis	entopeduncular nucleus, ventral part	ENv	92, 98, 107, 114; **22**; *18, 26, 31*	8
nucleus fasciculi longitudinalis medialis	nucleus of the medial longitudinal fascicle	NMLF	168, 173, 179; **46, 53**; *18, 26, 31*	2, 10, 16, 98, 100

Latin	English	Abbr.	Sect. No.	Text (page)
nucleus funiculi medialis	medial funicular nucleus	MFN	303, 319; **34**; *46, 58*	14, 16, 94, 95, 97
nucleus gustatorius secundarius	secondary gustatory nucleus	SGN	213; **22**; *26*	13, 15, 16, 95, 96
nucleus gustatorius tertius	tertiary gustatory nucleus (of Wullimann 88)	TGN	162, 168; *58*	10, 91, 95
nucleus habenularis dorsalis	dorsal habenular nucleus	Had	131, 136	9, 11
nucleus habenularis ventralis	ventral habenular nucleus	Hav	121, 125, 127, 131, 136	9, 11
nucleus intermedius thalami	intermediate thalamic nucleus	I	127	10, 90
nucleus interpeduncularis	interpeduncular nucleus	NIn	185, 196; **46, 53**; *46, 58*	11, 12, 13, 15, 16
nucleus isthmi	isthmic nucleus	NI	204, 208, 213; **9, 15**; *31*	15, 91, 92, 99, 100
nucleus lateralis hypothalami	lateral hypothalamic nucleus	LH	149, 153, 158; **34**	10
nucleus lateralis valvulae	nucleus lateralis valvulae	NLV	185, 196, 201, 204, 208; **9, 15, 22, 34**; *18, 26, 31*	15, 16, 91, 99
nucleus lemnisci lateralis	nucleus of the lateral lemniscus (of Prasada Rao et al. 87)	NLL	179, 185	12, 99
nucleus mesencephalicus nervi trigemini	mesencephalic nucleus of the trigeminal nerve	MNV	158	14, 97, 98
nucleus motorius nervi facialis	facial motor nucleus	NVIIm	251; **53**	14, 92, 93, 97, 98
nucleus motorius nervi trigemini, pars dorsalis	trigeminal motor nucleus, dorsal part	NVmd	208, 213, 219; **22**; *46*	13, 98
nucleus motorius nervi trigemini, pars ventralis	trigeminal motor nucleus, ventral part	NVmv	213, 219, 223; **22**; *58*	13, 98
nucleus motorius nervi vagi	vagal motor nucleus	NXm	279, 283, 290, 303; **46, 53**; *58*	14, 96, 98
nucleus nervi abducentis, pars caudalis	abducens nucleus, caudal part	NVIc	237, 239	14, 97
nucleus nervi oculomotorii	oculomotor nucleus	NIII	179, 185; **46, 53**; *31*	12, 13, 97, 100
nucleus nervi trochlearis	trochlear nucleus	NIV	196; **53**; *26, 31*	12, 13, 97
nucleus octavolateralis caudalis	caudal octavolateralis nucleus	CON	272, 279, 283; **22**; *31, 46*	14, 15, 92
nucleus octavolateralis medialis	medial octavolateralis nucleus	MON	230, 237, 239, 251, 260; **9, 15, 22, 34**; *31, 46*	14, 92, 93, 99
nucleus octavus anterior	anterior octaval nucleus	AON	223, 230; **15**; *46, 58*	14, 15, 93, 94, 97, 98

Latin	English	Abbr.	Sect. No.	Text (page)
nucleus octavus descendens	descending octaval nucleus	DON	239, 251, 260, 272; **22**, **46**, **53**;*46, 58, 76*	14, 93, 94, 97, 98, 99
nucleus octavus magnocellularis	magnocellular octaval nucleus	MaON	237, 239; **15**, **22**; *58*	14, 15, 93, 94, 98
nucleus octavus posterior	posterior octaval nucleus	PON	279; *76*	14, 93, 94
nucleus opticus accessorius dorsalis	dorsal accessory optic nucleus	DAO	127, 131, 136; **15**, **22**; *31*	11, 91, 97, 99
nucleus opticus accessorius ventralis	ventral accessory optic nucleus	VAO	136	11, 91, 99
nucleus paracommissuralis	paracommissural nucleus	PCN	149	10, 99
nucleus perilemniscularis	perilemniscal nucleus	PL	179, 185, 196; **9**; *31*	12, 99
nucleus periventricularis hypothalami, zona caudalis	caudal zone of periventricular hypothalamus	Hc	162, 168, 173; **34**, **46**, **53**; *76*	10
nucleus periventricularis hypothalami, zona dorsalis	dorsal zone of periventricular hypothalamus	Hd	149, 153, 158, 162, 168, 173, 179, 185; **22**, **34**; *76*	10
nucleus periventricularis hypothalami, zona ventralis	ventral zone of periventricular hypothalamus	Hv	127, 131, 136, 149, 153, 158; **46**, **53**; *58, 76*	10
nucleus periventricularis tuberculi posterioris	periventricular nucleus of posterior tuberculum	TPp	136, 149, 153, 158, 162; **34**, **46**, **53**; *31, 46*	10
nucleus posterior thalami	posterior thalamic nucleus	P	158, 162; *58*	10, 95
nucleus praeglomerulosus anterior	anterior preglomerular nucleus	PGa	136; **15**, **22**; *76*	10
nucleus praeglomerulosus caudalis	caudal preglomerular nucleus	PGc	173	10
nucleus praeglomerulosus lateralis	lateral preglomerular nucleus	PGl	136, 149, 153, 158; **9**, **15**, **22**; *58*	10, 92
nucleus praeglomerulosus medialis	medial preglomerular nucleus	PGm	149, 153, 158, 162, 168; **15**, **22**; *58*	10
nucleus praeopticus magnocellularis	magnocellular preoptic nucleus	PM	114; **46**; *26, 31*	9
nucleus praeopticus magnocellularis, pars gigantocellularis	gigantocellular part of magnocellular preoptic nucleus	PMg	121; **46**	9
nucleus praeopticus parvocellularis, pars anterior	parvocellular preoptic nucleus, anterior part	PPa	92, 98, 107; **34**, **46**, **53**; *18, 26, 31, 46*	9
nucleus praeopticus parvocellularis, pars posterior	parvocellular preoptic nucleus, posterior part	PPp	114, 121, 125, 127, 131; **46**, **53**; *31, 46*	9

Latin	English	Abbr.	Sect. No.	Text (page)
nucleus praetectalis accessorius	accessory pretectal nucleus (of Wullimann & Meyer 90)	APN	136; **22**; *18, 26, 31*	11
nucleus praetectalis centralis	central pretectal nucleus	CPN	127, 131, 136; **9, 15**; *18*	11, 90, 99, 100
nucleus praetectalis periventricularis, pars dorsalis	periventricular pretectal nucleus, dorsal part	PPd	149, 153; *18*	10, 91, 99, 100
nucleus praetectalis periventricularis, pars ventralis	periventricular pretectal nucleus, ventral part	PPv	149, 153, 158; **34 46, 53**; *18, 26*	10
nucleus praetectalis posterior	posterior prectectal nucleus (of Wullimann & Meyer 90)	PO	136; **22**; *18, 26, 31*	11
nucleus praetectalis superficialis, pars magnocellularis	magnocellular superficial pretectal nucleus	PSm	127, 131; **9, 15**; *26, 31*	11, 90, 91
nucleus praetectalis superficialis, pars parvocellularis	parvocellular superficial pretectal nucleus	PSp	121, 125, 127, 131; **9, 15**; *18, 26, 31*	11, 91
nucleus raphes inferior	inferior raphe	IR	272, 279, 283, 290	15
nucleus raphes superior	superior raphe	SR	201, 204, 208; **46**; *46, 58*	15, 98, 99
nucleus reticularis lateralis	lateral reticular nucleus	LRN	279	15, 99
nucleus rostrolateralis	rostrolateral nucleus (of Butler & Saidel 91)	R	125	10
nucleus ruber	red nucleus	NR	168	12, 16, 98, 99, 100
nucleus sensorius principalis nervi trigemini	primary sensory trigeminal nucleus	NⅤs	219	13, 16, 97, 98, 99
nucleus subglomerulosus	subglomerular nucleus	SG	168; **22**	10
nucleus suprachiasmaticus	suprachiasmatic nucleus	SC	114, 121, 125, 127; **53**; *46*	9, 91
nucleus taeniae	nucleus taeniae	NT	98, 107, 114; **9, 15**; *1*	8, 89
nucleus tangentialis	tangential nucleus	T	237, 239	14, 93, 94, 97, 98
nucleus tegmentalis dorsalis	dorsal tegmental nucleus	DTN	173, 179, 185, 196; **15, 22**; *18*	12, 15, 16, 99
nucleus tegmentalis rostralis	rostral tegmental nucleus (of Grover & Sharma 81)	RT	162, 168; **22**	12
nucleus tuberis anterior	anterior tuberal nucleus	ATN	136, 149, 153; **34**; *76*	10
nucleus tuberis posterior	posterior tuberal nucleus	PTN	149, 153, 158, 162; **34, 46, 53**; *58*	10, 11, 89, 101
nucleus ventrolateralis thalami	ventrolateral thalamic nucleus	VL	121, 125, 127, 131, 136; *18, 26*	10, 90, 96, 97, 100

Latin	English	Abbr.	Sect. No.	Text (page)
nucleus ventrolateralis tori semicircularis	ventrolateral nucleus of semicircular torus	TSvl	168, 173, 179, 185, 196, 201, 204, 208, 213; **1, 9**; *18, 26, 31*	12, 92, 100
nucleus ventromedialis thalami	ventromedial thalamic nucleus	VM	114, 121, 125, 127, 131, 136; *18, 26*	10, 93, 96, 97, 100
oliva inferior	inferior olive	IO	272, 279, 283, 290; **34**	15, 17, 99
organum paraventriculare	paraventricular organ	PVO	158	10
organum subcommissurale	subcommissural organ	SCO	149; *1*	10
populatio octavia secundaria	secondary octaval population (of McCormick & Hernandez 96)	SO	237, 239; **34, 53**; *46*	14, 93
radix caudalis nervi abducentis	caudal root of the abducens nerve	VIc	237	14
radix descendens nervi trigemini	descending trigeminal root	DV	223, 230, 237, 239, 251, 260, 272, 279, 283, 290, 303, 319; **22, 34, 46**; *58*	2, 13, 14, 16, 97
radix dorsalis	dorsal root	DR	363	15, 97
radix motoria nervi trigemini, pars dorsalis	dorsal motor root of the trigeminal nerve	Vmd	219	
radix motoria nervi trigemini, pars ventralis	ventral motor root of the trigeminal nerve	Vmv	223	
radix rostralis nervi abducentis	rostral root of the abducens nerve	VIr	223	14
radix sensoria nervi facialis	sensory root of the facial nerve	VIIs	230, 237, 239, 251, 260; **9, 15, 22, 34, 53**; *46*	14, 95
radix sensoria nervi trigemini	sensory root of the trigeminal nerve	Vs	219; **9, 15**; *46*	16, 97
recessus lateralis	lateral recess of diencephalic ventricle	LR	149, 153, 158, 162, 168, 173, 179, 185; **22, 34**; *76*	10
recessus posterior	posterior recess of diencephalic ventricle	PR	173; **34, 46, 53**	10
saccus dorsalis	dorsal sac	SD	121, 125, 127, 131; **46, 53**	9
stratum cellulare externum bulbi olfactorii	external cellular layer of olfactory bulb	ECL	23, 31, 50; **22, 34, 46, 53**; *1, 18, 26*	8
stratum cellulare internum bulbi olfactorii	internal cellular layer of olfactory bulb	ICL	31, 50; **22, 34, 46, 53**; *1, 18*	8

Latin	English	Abbr.	Sect. No.	Text (page)
stratum glomerulosum bulbi olfactorii	glomerular layer of olfactory bulb	GL	23, 31, 50; **15, 22, 34, 46, 53**; *1, 18, 26*	8, 89
stratum nervosum bulbi olfactorii	primary olfactory fiber layer	POF	23; **15, 22, 34, 46, 53**; *1, 18, 26*	8
stratum periventriculare tecti optici	periventricular gray zone of optic tectum	PGZ	136, 149, 153, 158, 162, 168, 173, 179, 185, 196, 201, 204, 208, 213, 219, 223; **1, 9, 15, 22, 34, 46, 53**; *1, 18, 26, 31, 46*	12
sulcus ypsiloniformis	ypsiloniform sulcus	SY	85, 92, 98	8
tectum opticum	optic tectum	TeO	127, 131, 136, 149, 153, 158, 162, 168, 173, 179, 185, 196, 201, 204, 208, 213, 219, 223, 230; **1, 9, 15, 22, 34, 46, 53**; *1, 18, 26, 31, 46, 58*	11, 12, 14, 16, 90, 99, 100
telencephalon	telencephalon	Tel		4, 7, 8, 9, 11, 89, 90, 93, 95, 97, 99, 100, 101
thalamus, pars dorsalis	dorsal thalamus	DT	46	4, 9, 10, 12, 90, 97, 101
thalamus, pars ventralis	ventral thalamus	VT	**34, 46, 53**	4, 9, 10, 97
torus lateralis	lateral torus	TLa	149, 153, 158, 162, 168, 173, 179, 185; **1, 9, 15**; *58, 76*	10
torus longitudinalis	longitudinal torus	TL	149, 153, 158, 162, 168, 173, 179, 185, 196, 201, 204; **34**; **46, 53**; *1*	12
torus semicircularis	semicircular torus	TS	162; **15**	12, 16, 92, 93, 96, 100
tractus habenulo-interpeduncularis = fasciculus retroflexus	habenulo-interpeduncular tract	FR	127, 131, 136, 149, 153, 158, 162, 168, 173, 179; **34, 53**; *1, 18, 26, 31, 46*	11
tractus bulbospinalis	bulbo-spinal tract	TBS	260, 272, 279, 283, 290, 303, 319; **22, 34, 46**	17, 98
tractus cerebellaris anterior	anterior cerebellar tract	AC	201, 204, 208; **22**; *18*	16
tractus cerebellaris posterior	posterior cerebellar tract	PC	219, 223; **22**; *18, 26, 31*	16

Latin	English	Abbr.	Sect. No.	Text (page)
tractus gustatorius secundarius	secondary gustatory tract	SGT	219, 223, 230, 237, 239, 251, 260, 272, 279, 283, 290, 303; **22**; *58*	14, 16, 95, 96
tractus mesencephalo-cerebellaris anterior	anterior mesencephalo-cerebellar tract	TMCa	179, 185, 196; **22**	16
tractus mesencephalo-cerebellaris posterior	posterior mesencephalo-cerebellar tract	TMCp	196; **22**	16
tractus olfactorius lateralis	lateral olfactory tract	LOT	50, 60, 71, 85, 92, 98; **9, 15**; *1*	7, 8, 9, 89
tractus olfactorius medialis	medial olfactory tract	MOT	50, 60, 71, 85, 92; **34, 46, 53**; *18*	7, 8, 9, 89
tractus opticus	optic tract	OT	107, 114, 121; **34**; *58*	11
tractus opticus dorsomedialis	dorsomedial optic tract	DOT	125, 127, 131, 136; **22**; *1, 18, 26, 31, 46*	11
tractus opticus ventrolateralis	ventrolateral optic tract	VOT	125, 127, 131, 136, 149, 153, 158, 162; **1, 9, 15, 22**; *46*	11
tractus praetectomamillaris	pretecto-mammillary tract	TPM	136, 149, 153, 158, 162, 168, 173, 179; **22**; *31, 46, 58, 76*	11
tractus tectobulbaris	tecto-bulbar tract	TTB	173, 179, 185, 196, 201, 204, 208; **1, 9, 15, 22**; *31, 46, 58*	16
tractus tectobulbaris cruciatus	crossed tecto-bulbar tract	TTBc	185, 196, 201, 204, 208, 213, 219, 223, 230, 237; **46, 53**; *58, 76*	16
tractus tectobulbaris rectus	uncrossed tecto-bulbar tract	TTBr	196, 201, 204, 208, 213, 219, 223, 230; **15**; *76*	16
tractus vestibulospinalis	vestibulo-spinal tract	TVS	272, 279, 283, 290; **34**	17, 98
valvula cerebelli, pars lateralis	lateral division of valvula cerebelli	Val	168, 173, 179, 185, 196, 201, 204, **15, 22, 34, 53**; *1*	13, 100
valvula cerebelli, pars medialis	medial division of valvula cerebelli	Vam	179, 185, 196, 201, 204, 208, 213; **34, 46, 53**; *1, 18, 26*	13
ventriculus diencephali	diencephalic ventricle	DiV	92, 98, 107, 114, 121, 125, 127, 131, 136, 149, 153, 158, 162, 168; **46, 53**; *18, 26, 31, 46, 58, 76*	9

Latin	English	Abbr.	Sect. No.	Text (page)
ventriculi telencephali	telencephalic ventricles	TelV	50, 60, 85, 98, 107, 114; *1, 18, 26*	8
ventriculus mesencephali	tectal ventricle	TeV	153, 158, 162, 168, 173, 179, 185, 196; **1, 9, 15, 22, 34, 46, 53**; *1, 18, 26*	12, 13, 14
ventriculus rhombencephali	rhombencephalic ventricle	RV	201, 204, 208, 213, 219, 223, 230, 237, 239, 251, 260, 272, 279, 283, 290, 303; **46, 53**; *18, 26, 31, 46, 58*	15, 16
zona limitans	zona limitans	ZL	136; **46**	

8 Index of English Terms

English	Latin	Abbr.	Sect. No.	Text (page)
central canal	canalis centralis	C	319, 363; **53**	15
central gray	griseum centrale	GC	196, 201, 204, 208, 213, 219, 223; **34**, **46**, **53**; *26, 31, 46*	12, 15
central nucleus of the inferior lobe	nucleus centralis lobi inferioris hypothalami	CIL	173, 179, 185; **15**, **22**; *76*	10, 95
central nucleus of semicircular torus	nucleus centralis tori semicircularis	TSc	168, 173, 179, 185, 196, 201, 204 208, 213; **9**; *1, 18*	12, 93
central nucleus of ventral telencephalic area	area ventralis telencephali, nucleus centralis	Vc	60, 71; **53**	8, 89
central posterior thalamic nucleus	nucleus centralis posterior thalami	CP	149, 153; **34**, **53**; *18, 26, 31*	9, 93
central pretectal nucleus	nucleus praetectalis centralis	CPN	127, 131, 136; **9**, **15**; *18*	11, 90, 99, 100
central zone of dorsal telencephalic area	area dorsalis telencephali, zona centralis	Dc	71, 85, 92, 98, 107; **15**, **22**, **34**, **46**, **53**; *1*	8, 92, 93, 95, 97, 100
cerebellar commissure	commissura cerebelli	Ccer	213, 219; **9**, **15**, **22**, **34**, **46**, **53**; *18*	13
cerebellar corpus	corpus cerebelli	CCe	204, 208, 213, 219, 223, 230, 237, 239, 251; **9**, **15**, **22**, **34**, **46**, **53**; *1, 18, 26*	13, 16, 92, 93, 99
cerebellar crest	crista cerebellaris	CC	237, 239, 251, 260, 272, 279; **15**, **22**, **34**, **46**, **53**; *18, 26, 31*	14
cerebellum	cerebellum	Ce	*31*	7, 13, 14, 15, 16, 91 92, 96, 99, 100
commissura infima of Haller	commissura infima Halleri	Cinf	319; **46**, **53**	15, 17
commissural nucleus of Cajal	nucleus commissuralis Cajal	NC	319; **46**, **53**; *46, 58*	14, 17, 94, 96
commissure of the posterior tuberculum	commissura tuberculi posterioris	Ctub	173; **46**, **53**	1, 11
commissure of the secondary gustatory nuclei	commissura nucleorum gustatoriorum secundariorum	Cgus	213; **34**, **46**, **53**; *26*	17
crossed tecto-bulbar tract	tractus tectobulbaris cruciatus	TTBc	185, 196, 201, 204, 208, 213, 219, 223, 230, 237; **46**; **53**; *58, 76*	16
descending octaval nucleus	nucleus octavus descendens	DON	239, 251, 260, 272; **22**, **46**, **53**; *46, 58, 76*	14, 93, 94, 97, 98, 99

English	Latin	Abbr.	Sect. No.	Text (page)
descending trigeminal root	radix descendens nervi trigemini	DV	223, 230, 237, 239, 251, 260, 272, 279, 283, 290, 303, 319; **22, 34, 46**; *58*	2, 13, 14, 16, 97
diencephalic ventricle	ventriculus diencephali	DiV	92, 98, 107, 114, 121, 125, 127, 131, 136, 149, 153, 158, 162, 168; **46, 53**; *18, 26, 31, 46, 58, 76*	9
diffuse nucleus of the inferior lobe	nucleus diffusus lobi inferioris hypothalami	DIL	149, 153, 158, 162, 168, 173, 179, 185, 196, 201, 204; **9, 15, 22, 34, 46, 53**; *76*	10
dorsal accessory optic nucleus	nucleus opticus accessorius dorsalis	DAO	127, 131, 136; **15, 22**; *31*	11, 91, 97, 99
dorsal funiculus	funiculus dorsalis	Fd	363	15, 16, 96
dorsal habenular nucleus	nucleus habenularis dorsalis	Had	131, 136	9, 11
dorsal horn	cornu dorsale	DH	363; **53**	15, 96
dorsal motor root of the trigeminal nerve	radix motoria nervi trigemini, pars dorsalis	Vmd	219	
dorsal nucleus of ventral telencephalic area	area ventralis telencephali, nucleus dorsalis	Vd	50, 60, 71, 85; **34, 46**; *1, 18*	8, 89, 101
dorsal part of lateral funiculus	funiculus lateralis, pars dorsalis	Fld	363; *58*	15, 16, 17
dorsal posterior thalamic nucleus	nucleus dorsalis posterior thalami	DP	149, 153; **34, 53**; *18, 26*	9, 90, 96, 97
dorsal root	radix dorsalis	DR	363	15, 97
dorsal sac	saccus dorsalis	SD	121, 125, 127, 131; **46, 53**	9
dorsal tegmental nucleus	nucleus tegmentalis dorsalis	DTN	173, 179, 185, 196; **15, 22**; *18*	12, 15, 16, 99
dorsal telencephalic area	area dorsalis telencephali	D	50, 60, 71, 85, 92, 98, 107, 114, 121, 125; **1, 9, 15, 22, 34, 46, 53**; *1, 18*	7, 8, 9, 89, 92, 97, 101
dorsal thalamus	thalamus, pars dorsalis	DT	**46**	4, 9, 10, 12, 90, 97, 101
dorsal zone of dorsal telencephalic area	area dorsalis telencephali, zona dorsalis	Dd	85, 92, 98; **22, 53**	8, 92, 97
dorsal zone of periventricular hypothalamus	nucleus periventricularis hypothalami, zona dorsalis	Hd	149, 153, 158, 162, 168, 173, 179, 185 **22, 34**; *76*	10

English	Latin	Abbr.	Sect. No.	Text (page)
dorsomedial optic tract	tractus opticus dorsomedialis	DOT	125, 127, 131, 136; **22**; *1, 18, 26, 31, 46*	11
Edinger-Westphal nucleus	nucleus Edinger-Westphal	EW	173	12
entopeduncular nucleus, dorsal part	nucleus entopeduncularis, pars dorsalis	ENd	85, 92, 98; **22**; *18*	8
entopeduncular nucleus, ventral part	nucleus entopeduncularis, pars ventralis	ENv	92, 98, 107, 114; **22**; *18, 26, 31*	8
epiphysis = pineal organ	epiphysis = corpus pineale	E	136; **46, 53**	6, 9
external cellular layer of olfactory bulb	stratum cellulare externum bulbi olfactorii	ECL	23, 31, 50; **22, 34, 46, 53**; *1, 18, 26*	8
facial lobe	lobus facialis	LVII	260, 272, 279, 283; **46, 53**; *46*	7, 14, 15, 94, 95, 98
facial motor nucleus	nucleus motorius nervi facialis	NVIIm	251; **53**	14, 92, 93, 97, 98
facial nerve	nervus facialis	VII	219, 223; **1**; *58*	13, 92, 94
gigantocellular part of magnocellular preoptic nucleus	nucleus praeopticus magnocellularis, pars gigantocellularis	PMg	121; **46**	9
glomerular layer of olfactory bulb	stratum glomerulosum bulbi olfactorii	GL	23, 31, 50; **15, 22, 34, 46, 53**; *1, 18, 26*	8, 89
glossopharyngeal lobe	lobus glossopharyngei	LIX	272	14, 94, 95
glossopharyngeal nerve	nervus glossopharyngeus	IX	272	13, 14, 94
granular eminence	eminentia granularis	EG	219, 223, 230, 237, 239; **1, 9, 15**; *1, 18, 26, 31*	13, 14, 16, 92, 93, 99
habenula	habenula	Ha	**34, 46, 53**; *1*	9, 11
habenular commissure	commissura habenularum	Chab	136; **46; 53**	1, 11
habenulo-interpeduncular tract	tractus habenulo-interpeduncularis = fasciculus retroflexus	FR	127, 131, 136, 149, 153, 158, 162, 168 173, 179; **34, 53**; *1, 18, 26, 31, 46*	11
horizontal commissure	commissura horizontalis	Chor	121, 125, 127, 131, 136, 149, 153, 158, 162; **15, 22, 34, 46, 53**; *31, 46, 58*	11
hypophysis = pituitary	hypophysis = glandula pituitaria	Pit	153, 158, 162, 168, 173	6, 10
inferior lobe of hypothalamus	lobus inferior hypothalami	IL		10, 12, 91, 95
inferior olive	oliva inferior	IO	272, 279, 283, 290; **34**	15, 17, 99

English	Latin	Abbr.	Sect. No.	Text (page)
inferior raphe	nucleus raphes inferior	IR	272, 279, 283, 290	15
inferior reticular formation	formatio reticularis, pars inferior	IRF	272, 279, 283, 290, 303, 319; **34, 46**; *76*	15, 16, 17, 98
inner arcuate fibers	fibrae arcuatae internae	IAF	237, 239, 279; *58*	14, 16
intermediate reticular formation	formatio reticularis, pars intermedia	IMRF	230, 237, 239, 251, 260; **34, 46, 53**; *76*	14, 15, 17, 98
intermediate thalamic nucleus	nucleus intermedius thalami	I	127	10, 90
internal cellular layer of olfactory bulb	stratum cellulare internum bulbi olfactorii	ICL	31, 50; **22, 34, 46, 53**; *1, 18*	8
interpeduncular nucleus	nucleus interpeduncularis	NIn	185, 196; **46, 53**; *46, 58*	11, 12, 13, 15, 16
intertectal commissure = tectal commissure	commissura intertectalis = commissura tecti	Ctec	149, 153, 158, 162, 168, 173, 179, 185, 196, 201; **34, 46, 53**	1, 12
isthmic nucleus	nucleus isthmi	NI	204, 208, 213; **9, 15**; *31*	15, 91, 92, 99, 100
lateral division of valvula cerebelli	valvula cerebelli, pars lateralis	Val	168, 173, 179, 185, 196, 201, 204; **15, 22, 34, 53**; *1*	13, 100
lateral forebrain bundle	fasciculus lateralis telencephali	LFB	92, 98, 107, 114, 121, 125, 127, 131, 136, 149; **15, 22, 34**; *1, 18, 26*	8, 9, 11
lateral hypothalamic nucleus	nucleus lateralis hypothalami	LH	149, 153, 158; **34**	10
lateral lemniscus = lateral longitudinal fascicle	lemniscus lateralis = fasciculus longitudinalis lateralis	LLF	168, 173, 179, 185, 196, 201, 204, 208, 213, 219, 223; **9, 15, 22, 34**; *18, 26, 31, 46, 58*	13, 16, 92, 93
lateral line nerves	nervi lineae lateralis	LL		13, 14, 91, 92, 93
lateral longitudinal fascicle = lateral lemniscus	fasciculus longitudinalis lateralis = lemniscus lateralis	LLF	168, 173, 179, 185, 196, 201, 204, 208, 213, 219, 223; **9, 15, 22, 34**; *18, 26 31, 46, 58*	13, 16, 92, 93
lateral nucleus of ventral telencephalic area	area ventralis telencephali, nucleus lateralis	Vl	60, 71; **15, 22, 53**; *18, 26, 31*	8
lateral olfactory tract	tractus olfactorius lateralis	LOT	50, 60, 71, 85, 92, 98; **9, 15**; *1*	7, 8, 9, 89
lateral preglomerular nucleus	nucleus praeglomerulosus lateralis	PGl	136, 149, 153, 158; **9, 15, 22**; *58*	10, 92

English	Latin	Abbr.	Sect. No.	Text (page)
lateral recess of diencephalic ventricle	recessus lateralis	LR	149, 153, 158, 162, 168, 173, 179, 185; **22**, **34**; *76*	10
lateral reticular nucleus	nucleus reticularis lateralis	LRN	279	15, 99
lateral torus	torus lateralis	TLa	149, 153, 158, 162, 168, 173, 179, 185; **1**; **9**, **15**; *58, 76*	10
lateral zone of dorsal telencephalic area	area dorsalis telencephali, zona lateralis	Dl	71, 85, 92, 98, 107, 114; **1**, **9**, **15**, **22**; *1*	8, 92
locus coeruleus	locus coeruleus	LC	204	15, 99, 101
longitudinal torus	torus longitudinalis	TL	149, 153, 158, 162, 168, 173, 179, 185, 196, 201, 204; **34**, **46**, **53**; *1*	12
magnocellular octaval nucleus	nucleus octavus magnocellularis	MaON	237, 239; **15**, **22**; *58*	14, 15, 93, 94, 98
magnocellular preoptic nucleus	nucleus praeopticus magnocellularis	PM	114; **46**; *26, 31*	9
magnocellular superficial pretectal nucleus	nucleus praetectalis superficialis, pars magnocellularis	PSm	127, 131; **9**, **15**; *26, 31*	11, 90, 91
mammillary body	corpus mamillare	CM	173, 179, 185; **34**, **46**, **53**; *58, 76*	10, 11, 91
Mauthner axon	fibra Mauthneri	MA	237, 239, 251, 260, 272, 279, 283, 290, 303, 319, 363; **46**	15, 16
Mauthner cell	cellula Mauthneri	MAC	230; *58*	15, 98, 99
medial division of valvula cerebelli	valvula cerebelli, pars medialis	Vam	179, 185, 196, 201, 204, 208, 213; **34**, **46**, **53**; *1, 18, 26*	13
medial forebrain bundle	fasciculus medialis telencephali	MFB	98, 107, 114, 121, 125, 127, 131, 136, 149; **34**, **53**; *18*	9, 11
medial funicular nucleus	nucleus funiculi medialis	MFN	303, 319; **34**; *46, 58*	14, 16, 94, 95, 97
medial longitudinal fascicle	fasciculus longitudinalis medialis	MLF	173, 179, 185, 196, 201, 204, 208, 213, 219, 223, 230, 237, 239, 251, 260, 272, 279, 283, 290, 303, 319, 363; **46**, **53**; *26, 31, 46, 58, 76*	2, 15, 16, 98
medial octavolateralis nucleus	nucleus octavolateralis medialis	MON	230, 237, 239, 251, 260; **9**, **15**, **22**, **34**; *31, 46*	14, 92, 93, 99

English	Latin	Abbr.	Sect. No.	Text (page)
medial olfactory tract	tractus olfactorius medialis	MOT	50, 60, 71, 85, 92; **34, 46, 53**; *18*	7, 8, 9, 89
medial preglomerular nucleus	nucleus praeglomerulosus medialis	PGm	149, 153, 158, 162, 168; **15, 22**; *58*	10
medial zone of dorsal telencephalic area	area dorsalis telencephali, zona medialis	Dm	71, 85, 92, 98, 107, 114; **34, 46, 53**; *1*	8, 92, 93, 95, 97
medulla oblongata	medulla oblongata	MO		12, 13, 14, 16, 17, 96 97, 99
mesencephalic nucleus of the trigeminal nerve	nucleus mesencephalicus nervi trigemini	MNV	158	14, 97, 98
nucleus lateralis valvulae	nucleus lateralis valvulae	NLV	185, 196, 201, 204, 208; **9, 15, 22, 34**; *18, 26, 31*	15, 16, 91, 99
nucleus of the descending trigeminal root	nucleus descendens nervi trigemini	NDV	239, 251, 260	13, 97, 98, 99
nucleus of the lateral lemniscus (of Prasada Rao et al. 87)	nucleus lemnisci lateralis	NLL	179, 185	12, 99
nucleus of the medial longitudinal fascicle	nucleus fasciculi longitudinalis medialis	NMLF	168, 173, 179; **46, 53**; *18, 26, 31*	2, 10, 16, 98, 100
nucleus taeniae	nucleus taeniae	NT	98, 107, 114; **9, 15**; *1*	8, 89
octaval nerve	nervus octavus	VIII	230, 237, 239, 251, 260, 272; **9, 15**; *58, 76*	13, 14, 93, 94
oculomotor nerve	nervus oculomotorius	III	179; **1, 9, 15, 22, 34, 46, 53**; *46, 58*	12
oculomotor nucleus	nucleus nervi oculomotorii	NIII	179, 185; **46, 53**; *31*	12, 13, 97, 100
olfactory bulb	bulbus olfactorius	OB		7, 8, 89
olfactory nerve	nervus olfactorius	I		7, 89
optic chiasm	chiasma opticum	CO	107, 114; **53**; *58*	11
optic nerve	nervus opticus (=II)	ON	107; **15, 22, 34, 46, 53**; *58*	11, 90
optic tectum	tectum opticum	TeO	127, 131, 136, 149, 153, 158, 162, 168, 173, 179, 185, 196, 201, 204, 208, 213, 219, 223, 230; **1, 9, 15, 22, 34, 46, 53**; *1, 18, 26, 31, 46, 58*	11, 12, 14, 16, 90, 99 100
optic tract	tractus opticus	OT	107, 114, 121; **34**; *58*	11
paracommissural nucleus	nucleus paracommissuralis	PCN	149	10, 99

English	Latin	Abbr.	Sect. No.	Text (page)
paraventricular organ	organum paraventriculare	PVO	158	10
parvocellular preoptic nucleus, anterior part	nucleus praeopticus parvocellularis, pars anterior	PPa	92, 98, 107; **34, 46, 53**; *18, 26, 31, 46*	9
parvocellular preoptic nucleus, posterior part	nucleus praeopticus parvocellularis, pars posterior	PPp	114, 121, 125, 127, 131; **46, 53**; *31, 46*	9
parvocellular superficial pretectal nucleus	nucleus praetectalis superficialis, pars parvocellularis	PSp	121, 125, 127, 131; **9, 15**; *18, 26, 31*	11, 91
perilemniscal nucleus	nucleus perilemniscularis	PL	179, 185, 196; **9**; *31*	12, 99
periventricular gray zone of optic tectum	stratum periventriculare tecti optici	PGZ	136, 149, 153, 158, 162, 168, 173, 179, 185, 196, 201, 204, 208; 213; 219; 223; **1, 9, 15, 22, 34, 46, 53**; *1, 18, 26, 31, 46*	12
periventricular nucleus of posterior tuberculum	nucleus periventricularis tuberculi posterioris	TPp	136, 149, 153, 158, 162; **34, 46, 53**; *31, 46*	10
periventricular pretectal nucleus, dorsal part	nucleus praetectalis periventricularis, pars dorsalis	PPd	149, 153; *18*	10, 91, 99, 100
periventricular pretectal nucleus, ventral part	nucleus praetectalis periventricularis, pars ventralis	PPv	149, 153, 158; **34, 46, 53**; *18, 26*	10
pineal organ = epiphysis	corpus pineale = epiphysis	E	136; **46, 53**	6, 9
pituitary = hypophysis	glandula pituitaria = hypophysis	Pit	153, 158, 162, 168, 173	6, 10
postcommissural nucleus of ventral telencephalic area	area ventralis telencephali, nucleus postcommissuralis	Vp	98, 107; **34, 46**; *1*	8
posterior cerebellar tract	tractus cerebellaris posterior	PC	219, 223; **22**; *18, 26, 31*	16
posterior commissure	commissura posterior	Cpost	149, 153, 158, 162, 168; **22, 34, 46, 53**; *1, 18*	10, 11
posterior lateral line nerve	nervus lineae lateralis posterioris	PLLN	251, 260, 272; **15**; *46, 58*	13, 92
posterior mesencephalo-cerebellar tract	tractus mesencephalo-cerebellaris posterior	TMCp	196; **22**	16
posterior octaval nucleus	nucleus octavus posterior	PON	279; *76*	14, 93, 94
posterior prectectal nucleus (of Wullimann & Meyer 90)	nucleus praetectalis posterior	PO	136; **22**; *18, 26, 31*	11
posterior recess of diencephalic ventricle	recessus posterior	PR	173; **34, 46, 53**	10
posterior thalamic nucleus	nucleus posterior thalami	P	158, 162; *58*	10, 95

English	Latin	Abbr.	Sect. No.	Text (page)
posterior tuberal nucleus	nucleus tuberis posterior	PTN	149, 153, 158, 162; **34, 46, 53**; *58*	10, 11, 89, 101
posterior zone of dorsal telencephalic area	area dorsalis telencephali, zona posterior	Dp	71, 85, 92, 98, 107, 114; **1, 9, 15, 22, 34, 46, 53**; *1, 18*	8, 9, 89, 101
postoptic commissure = supraoptic commissure	commissura postoptica = commissura supraoptica	Cpop	121, 125, 127; **34, 46, 53**; *58*	3, 11
preglomerular area	area praeglomerulosa	PG		11, 90, 95, 96
pretecto-mammillary tract	tractus praetectomamillaris	TPM	136, 149, 153, 158, 162, 168, 173, 179; **22**; *31, 46, 58, 76*	11
primary olfactory fiber layer	stratum nervosum bulbi olfactorii	POF	23; **15, 22, 34, 46, 53**; *1, 18, 26*	8
primary sensory trigeminal nucleus	nucleus sensorius principalis nervi trigemini	NVs	219	13, 16, 97, 98, 99
red nucleus	nucleus ruber	NR	168	12, 16, 98, 99, 100
rhombencephalic ventricle	ventriculus rhombencephali	RV	201, 204, 208, 213, 219, 223, 230, 237, 239, 251, 260, 272, 279, 283, 290, 303; **46, 53**; *18, 26, 31, 46, 58*	15, 16
rostral root of the abducens nerve	radix rostralis nervi abducentis	VIr	223	14
rostral tegmental nucleus (of Grover & Sharma 81)	nucleus tegmentalis rostralis	RT	162, 168; **22**	12
rostrolateral nucleus (of Butler & Saidel 91)	nucleus rostrolateralis	R	125	10
secondary gustatory nucleus	nucleus gustatorius secundarius	SGN	213; **22**; *26*	13, 15, 16, 95, 96
secondary gustatory tract	tractus gustatorius secundarius	SGT	219, 223, 230, 237, 239, 251, 260, 272, 279, 283, 290, 303; **22**; *58*	14, 16, 95, 96
secondary octaval population (of McCormick & Hernandez 96)	populatio octavia secundaria	SO	237, 239; **34, 53**; *46*	14, 93
semicircular torus	torus semicircularis	TS	162; **15**	12, 16, 92, 93, 96, 100
sensory root of the facial nerve	radix sensoria nervi facialis	VIIs	230, 237, 239, 251, 260; **9, 15, 22, 34, 53**; *46*	14, 95
sensory root of the trigeminal nerve	radix sensoria nervi trigemini	Vs	219; **9, 15**; *46*	16, 97

English	Latin	Abbr.	Sect. No.	Text (page)
spinal cord	medulla spinalis	MS		1, 3, 4, 7, 12, 15, 16, 17, 94, 96, 97, 98, 99
subcommissural organ	organum subcommissurale	SCO	149; *1*	10
subglomerular nucleus	nucleus subglomerulosus	SG	168; **22**	10
superior raphe	nucleus raphes superior	SR	201, 204, 208; **46**; *46, 58*	15, 98, 99
superior reticular formation	formatio reticularis, pars superior	SRF	196, 201, 204, 208, 213, 219, 223; **34, 46, 53**; *46, 58, 76*	12, 14, 15, 16, 98
suprachiasmatic nucleus	nucleus suprachiasmaticus	SC	114, 121, 125, 127; **53**; *46*	9, 91
supracommissural nucleus of ventral telencephalic area	area ventralis telencephali, nucleus supracommissuralis	Vs	92; **34, 46, 53**; *1*	8, 9
supraoptic commissure = postoptic commissure	commissura supraoptica = commissura postoptica	Cpop	121, 125, 127; **34, 46, 53**; *58*	3, 11
tangential nucleus	nucleus tangentialis	T	237, 239	14, 93, 94, 97, 98
tectal commissure = intertectal commissure	commissura tecti = commissura intertectalis	Ctec	149, 153, 158, 162, 168, 173, 179, 185 196, 201; **34, 46, 53**	1, 12
tectal ventricle	ventriculus mesencephali	TeV	153, 158, 162, 168, 173, 179, 185, 196; **1, 9, 15, 22, 34, 46, 53**; *1, 18, 26*	12, 13, 14
tecto-bulbar tract	tractus tectobulbaris	TTB	173, 179, 185, 196, 201, 204, 208; **1, 9, 15, 22**; *31, 46, 58*	16
telencephalic ventricles	ventriculi telencephali	TelV	50, 60, 85, 98, 107, 114; *1, 18, 26*	8
telencephalon	telencephalon	Tel		4, 7, 8, 9, 11, 89, 90, 93, 95, 97, 99, 100, 101
tertiary gustatory nucleus (of Wullimann 88)	nucleus gustatorius tertius	TGN	162, 168; *58*	10, 91, 95
trigeminal motor nucleus, dorsal part	nucleus motorius nervi trigemini, pars dorsalis	NⅤmd	208, 213, 219; **22**; *46*	13, 98
trigeminal motor nucleus, ventral part	nucleus motorius nervi trigemini, pars ventralis	NⅤmv	213, 219, 223; **22**; *58*	13, 98
trigeminal nerve	nervus trigeminus	Ⅴ	213; **1**; *58*	13, 14, 95, 97
trochlear decussation	decussatio trochlearis	DⅣ	204; **34, 46, 53**	12
trochlear nerve	nervus trochlearis	Ⅳ	201, 208, 213; **1, 9, 15, 22**; *26, 31*	12, 92

English	Latin	Abbr.	Sect. No.	Text (page)
trochlear nucleus	nucleus nervi trochlearis	NIV	196; **53**; *26, 31*	12, 13, 97
tuberal hypothalamus	hypothalamus, pars tuberalis	TH		10, 93
uncrossed tecto-bulbar tract	tractus tectobulbaris rectus	TTBr	196, 201, 204, 208, 213, 219, 223, 230; **15**; *76*	16
vagal lobe	lobus vagi	LX	279, 283, 290, 303; **22, 34, 46, 53**; *26, 31, 46*	7, 14, 15, 92, 94, 95, 96
vagal motor nucleus	nucleus motorius nervi vagi	NXm	279, 283, 290, 303; **46, 53**; *58*	14, 96, 98
vagal nerve	nervus vagus	X	279, 283, 290; *46, 58*	13, 14, 92, 94, 95, 96
vascular lacuna of areae postrema	lacuna vasculosa area postremae	Vas	168, 173, 179; **46**	
ventral accessory optic nucleus	nucleus opticus accessorius ventralis	VAO	136	11, 91, 99
ventral funiculus	funiculus ventralis	Fv	303, 319, 363	16
ventral habenular nucleus	nucleus habenularis ventralis	Hav	121, 125, 127, 131, 136	9, 11
ventral horn	cornu ventrale	VH	363	15
ventral motor root of the trigeminal nerve	radix motoria nervi trigemini, pars ventralis	Vmv	223	
ventral nucleus of ventral telencephalic area	area ventralis telencephali, nucleus ventralis	Vv	60, 71, 85; **34, 46, 53**; *18, 26, 31*	8, 9, 89
ventral part of lateral funiculus	funiculus lateralis, pars ventralis	Flv	303, 319, 363	15, 16, 17
ventral rhombencephalic commissure	commissura ventralis rhombencephali	Cven	201, 204, 208, 213, 219, 223, 230, 237, 239, 251, 272; **34, 46, 53**; *46, 58, 76*	14, 15, 16
ventral telencephalic area	area ventralis telencephali	V	50, 60, 71, 85, 92, 98, 107; **15, 22, 34, 46, 53**; *1, 18, 26, 31*	7, 8, 9, 89, 93, 101
ventral thalamus	thalamus, pars ventralis	VT	**34, 46, 53**	4, 9, 10, 97
ventral zone of periventricular hypothalamus	nucleus periventricularis hypothalami, zona ventralis	Hv	127, 131, 136, 149, 153, 158; **46, 53**; *58, 76*	10
ventrolateral nucleus of semicircular torus	nucleus ventrolateralis tori semicircularis	TSvl	168, 173, 179, 185, 196, 201, 204, 208, 213; **1, 9**; *18, 26, 31*	12, 92, 100
ventrolateral optic tract	tractus opticus ventrolateralis	VOT	125, 127, 131, 136, 149, 153, 158, 162; **1, 9, 15, 22**; *46*	11

English	Latin	Abbr.	Sect. No.	Text (page)
ventrolateral thalamic nucleus	nucleus ventrolateralis thalami	VL	121, 125, 127, 131, 136; *18, 26*	10, 90, 96, 97, 100
ventromedial thalamic nucleus	nucleus ventromedialis thalami	VM	114, 121, 125, 127, 131, 136; *18, 26*	10, 93, 96, 97, 100
vestibulo-spinal tract	tractus vestibulospinalis	TVS	272, 279, 283, 290; **34**	17, 98
ypsiloniform sulcus	sulcus ypsiloniformis	SY	85, 92, 98	8
zona limitans	zona limitans	ZL	136, **46**	

9 Index of Abbreviations

Abbr.	Latin	English	Sect. No.	Text (page)
A	nucleus anterior thalami	anterior thalamic nucleus	131, 136; **34**, **53**; *18*	9, 90, 100
AC	tractus cerebellaris anterior	anterior cerebellar tract	201, 204, 208; **22**; *18*	16
ALLN	nervi lineae lateralis anterioris	anterior lateral line nerves	219, 223, 230, 237, 239; **1, 9, 15**; *46, 58*	13, 14, 92
AON	nucleus octavus anterior	anterior octaval nucleus	223, 230; **15**; *46, 58*	14, 15, 93, 94, 97, 98
APN	nucleus praetectalis accessorius	accessory pretectal nucleus (of Wullimann & Meyer 90)	136; **22**; *18, 26; 31*	11
ATN	nucleus tuberis anterior	anterior tuberal nucleus	136, 149, 153; **34**; *76*	10
C	canalis centralis	central canal	319, 363; **53**	15
Cans	commissura ansulata	ansulate commissure	179; **34, 46, 53**; *58*	2, 16
Cantd	commissura anterior, pars dorsalis	anterior commissure, dorsal part	85, 92; **34, 46, 53**	9
Cantv	commissura anterior, pars ventralis	anterior commissure, ventral part	92; **34, 46, 53**; *18*	9
CC	crista cerebellaris	cerebellar crest	237, 239, 251, 260, 272, 279; **15, 22, 34, 46, 53**; *18, 26, 31*	14
CCe	corpus cerebelli	cerebellar corpus	204, 208, 213, 219, 223, 230, 237, 239, 251; **9, 15, 22, 34, 46, 53**; *1, 18, 26*	13, 16, 92, 93, 99
Ccer	commissura cerebelli	cerebellar commissure	213, 219; **9, 15, 22, 34, 46, 53**; *18*	13
Ce	cerebellum	cerebellum	*31*	7, 13, 14, 15, 16, 91, 92, 96, 99, 100
Cgus	commissura nucleorum gustatoriorum secundariorum	commissure of the secondary gustatory nuclei	213; **34, 46, 53**; *26*	17
Chab	commissura habenularum	habenular commissure	136; **46, 53**	1, 11
Chor	commissura horizontalis	horizontal commissure	121, 125, 127, 131, 136, 149, 153, 158, 162; **15, 22, 34, 46, 53**; *31, 46, 58*	11
CIL	nucleus centralis lobi inferioris hypothalami	central nucleus of the inferior lobe	173, 179, 185; **15, 22**; *76*	10, 95

127

Abbr.	Latin	English	Sect. No.	Text (page)
Cinf	commissura infima Halleri	commissura infima of Haller	319; **46, 53**	15, 17
CM	corpus mamillare	mammillary body	173, 179, 185; **34, 46, 53**; *58, 76*	10, 11, 91
CO	chiasma opticum	optic chiasm	107, 114; **53**; *58*	11
CON	nucleus octavolateralis caudalis	caudal octavolateralis nucleus	272, 279, 283; **22**; *31, 46*	14, 15, 92
CP	nucleus centralis posterior thalami	central posterior thalamic nucleus	149, 153; **34, 53**, *18, 26, 31*	9, 93
CPN	nucleus praetectalis centralis	central pretectal nucleus	127, 131, 136; **9, 15**; *18*	11, 90, 99, 100
Cpop	commissura supraoptica = commissura postoptica	supraoptic commissure = postoptic commissure	121, 125, 127; **34, 46, 53**; *58*	3, 11
Cpost	commissura posterior	posterior commissure	149, 153, 158, 162, 168; **22, 34, 46, 53**; *1, 18*	10, 11
Ctec	commissura tecti = commissura intertectalis	tectal commissure = intertectal commissure	149, 153, 158, 162, 168, 173, 179, 185, 196, 201; **34, 46, 53**	1, 12
Ctub	commissura tuberculi posterioris	commissure of the posterior tuberculum	173; **46, 53**	1, 11
Cven	commissura ventralis rhombencephali	ventral rhombencephalic commissure	201, 204, 208, 213, 219, 223, 230, 237, 239, 251, 272; **34, 46, 53**; *46, 58, 76*	14, 15, 16
D	area dorsalis telencephali	dorsal telencephalic area	50, 60, 71, 85, 92, 98, 107, 114, 121, 125; **1, 9, 15, 22, 34, 46, 53**; *1, 18*	7, 8, 9, 89, 92, 97, 101
DAO	nucleus opticus accessorius dorsalis	dorsal accessory optic nucleus	127, 131, 136; **15, 22**; *31*	11, 91, 97, 99
Dc	area dorsalis telencephali, zona centralis	central zone of dorsal telencephalic area	71, 85, 92, 98, 107; **15, 22, 34, 46, 53**; *1*	8, 92, 93, 95, 97, 100
Dd	area dorsalis telencephali, zona dorsalis	dorsal zone of dorsal telencephalic area	85, 92, 98; **22, 53**	8, 92, 97
DH	cornu dorsale	dorsal horn	363; **53**	15, 96
DIL	nucleus diffusus lobi inferioris hypothalami	diffuse nucleus of the inferior lobe	149, 153, 158, 162, 168, 173, 179, 185, 196, 201, 204; **9, 15, 22, 34, 46, 53**; *76*	10
DiV	ventriculus diencephali	diencephalic ventricle	92, 98, 107, 114, 121, 125, 127, 131, 136, 149, 153, 158, 162, 168; **46, 53**; *18, 26, 31, 46, 58, 76*	9

Abbr.	Latin	English	Sect. No.	Text (page)
Dl	area dorsalis telencephali, zona lateralis	lateral zone of dorsal telencephalic area	71, 85, 92, 98, 107, 114; *1, 9, 15, 22; 1*	8, 92
Dm	area dorsalis telencephali, zona medialis	medial zone of dorsal telencephalic area	71, 85, 92, 98, 107, 114; *34, 46, 53; 1*	8, 92, 93, 95, 97
DON	nucleus octavus descendens	descending octaval nucleus	239, 251, 260, 272; *22, 46, 53; 46, 58, 76*	14, 93, 94, 97, 98, 99
DOT	tractus opticus dorsomedialis	dorsomedial optic tract	125, 127, 131, 136; *22; 1, 18, 26, 31, 46*	11
Dp	area dorsalis telencephali, zona posterior	posterior zone of dorsal telencephalic area	71, 85, 92, 98, 107, 114; *1, 9, 15, 22, 34, 46, 53; 1, 18*	8, 9, 89, 101
DP	nucleus dorsalis posterior thalami	dorsal posterior thalamic nucleus	149, 153; *34, 53; 18, 26*	9, 90, 96, 97
DR	radix dorsalis	dorsal root	363	15, 97
DT	thalamus, pars dorsalis	dorsal thalamus	**46**	4, 9, 10, 12, 90, 97, 101
DTN	nucleus tegmentalis dorsalis	dorsal tegmental nucleus	173, 179, 185, 196; *15, 22; 18*	12, 15, 16, 99
DIV	decussatio trochlearis	trochlear decussation	204; *34, 46, 53*	12
DV	radix descendens nervi trigemini	descending trigeminal root	223, 230, 237, 239, 251, 260, 272, 279, 283, 290, 303, 319; *22, 34, 46; 58*	2, 13, 14, 16, 97
E	epiphysis = corpus pineale	epiphysis = pineal organ	136; *46, 53*	6, 9
ECL	stratum cellulare externum bulbi olfactorii	external cellular layer of olfactory bulb	23, 31, 50; *22, 34, 46, 53; 1, 18, 26*	8
EG	eminentia granularis	granular eminence	219, 223, 230, 237, 239; *1, 9, 15; 1, 18, 26, 31*	13, 14, 16, 92, 93, 99
ENd	nucleus entopeduncularis, pars dorsalis	entopeduncular nucleus, dorsal part	85, 92, 98; *22; 18*	8
ENv	nucleus entopeduncularis, pars ventralis	entopeduncular nucleus, ventral part	92, 98, 107, 114; *22; 18, 26, 31*	8
EW	nucleus Edinger-Westphal	Edinger-Westphal nucleus	173	12
Fd	funiculus dorsalis	dorsal funiculus	363	15, 16, 96
Fld	funiculus lateralis, pars dorsalis	dorsal part of lateral funiculus	363; *58*	15, 16, 17
Flv	funiculus lateralis, pars ventralis	ventral part of lateral funiculus	303, 319, 363	15, 16, 17
FR	fasciculus retroflexus = tractus habenulointerpeduncularis	habenulo-interpeduncular tract	127, 131, 136, 149, 153, 158, 162, 168, 173, 179; *34, 53; 1, 18, 26, 31, 46*	11

Abbr.	Latin	English	Sect. No.	Text (page)
Fv	funiculus ventralis	ventral funiculus	303, 319, 363	16
GC	griseum centrale	central gray	196, 201, 204, 208, 213, 219, 223; **34, 46, 53**; *26, 31, 46*	12, 15
GL	stratum glomerulosum bulbi olfactorii	glomerular layer of olfactory bulb	23, 31, 50; **15, 22, 34, 46, 53**; *1, 18, 26*	8, 89
Ha	habenula	habenula	**34, 46, 53**; *1*	9, 11
Had	nucleus habenularis dorsalis	dorsal habenular nucleus	131, 136	9, 11
Hav	nucleus habenularis ventralis	ventral habenular nucleus	121, 125, 127, 131, 136	9, 11
Hc	nucleus periventricularis hypothalami, zona caudalis	caudal zone of periventricular hypothalamus	162, 168, 173; **34, 46, 53**; *76*	10
Hd	nucleus periventricularis hypothalami, zona dorsalis	dorsal zone of periventricular hypothalamus	149, 153, 158, 162, 168, 173, 179, 185; **22, 34**; *76*	10
Hv	nucleus periventricularis hypothalami, zona ventralis	ventral zone of periventricular hypothalamus	127, 131, 136, 149, 153, 158; **46, 53**; *58, 76*	10
I	nucleus intermedius thalami	intermediate thalamic nucleus	127	10, 90
IAF	fibrae arcuatae internae	inner arcuate fibers	237, 239, 279; *58*	14, 16
ICL	stratum cellulare internum bulbi olfactorii	internal cellular layer of olfactory bulb	31, 50; **22, 34, 46, 53**; *1, 18*	8
IL	lobus inferior hypothalami	inferior lobe of hypothalamus		10, 12, 91, 95
IMRF	formatio reticularis, pars intermedia	intermediate reticular formation	230, 237, 239, 251, 260; **34, 46, 53**; *76*	14, 15, 17, 98
IO	oliva inferior	inferior olive	272, 279, 283, 290; **34**	15, 17, 99
IR	nucleus raphes inferior	inferior raphe	272, 279, 283, 290	15
IRF	formatio reticularis, pars inferior	inferior reticular formation	272, 279, 283, 290, 303, 319; **34, 46**; *76*	15, 16, 17, 98
LC	locus coeruleus	locus coeruleus	204	15, 99, 101
LCa	lobus caudalis cerebelli	caudal lobe of cerebellum	230, 237, 239, 251, 260; **22, 34, 46, 53**; *1, 18, 26*	13, 99
LFB	fasciculus lateralis telencephali	lateral forebrain bundle	92, 98, 107, 114, 121, 125, 127, 131, 136, 149; **15, 22, 34**; *1, 18, 26*	8, 9, 11
LH	nucleus lateralis hypothalami	lateral hypothalamic nucleus	149, 153, 158; **34**	10
LL	nervi lineae lateralis	lateral line nerves		13, 14, 91, 92, 93

Abbr.	Latin	English	Sect. No.	Text (page)
LLF	fasciculus longitudinalis lateralis = lemniscus lateralis	lateral longitudinal fascicle = lateral lemniscus	168, 173, 179, 185, 196, 201, 204, 208, 213, 219, 223; **9**, **15, 22, 34**; *18, 26, 31, 46, 58*	13, 16, 92, 93
LOT	tractus olfactorius lateralis	lateral olfactory tract	50, 60, 71, 85, 92, 98; **9, 15**; *1*	7, 8, 9, 89
LR	recessus lateralis	lateral recess of diencephalic ventricle	149, 153, 158, 162, 168, 173, 179, 185; **22, 34**; *76*	10
LRN	nucleus reticularis lateralis	lateral reticular nucleus	279	15, 99
LVII	lobus facialis	facial lobe	260, 272, 279, 283; **46, 53**; *46*	7, 14, 15, 94, 95, 98
LIX	lobus glossopharyngei	glossopharyngeal lobe	272	14, 94, 95
LX	lobus vagi	vagal lobe	279, 283, 290, 303; **22, 34, 46, 53**; *26, 31, 46*	7, 14, 15, 92, 94, 95, 96
MA	fibra Mauthneri	Mauthner axon	237, 239, 251, 260, 272, 279, 283, 290, 303, 319, 363; **46**	15, 16
MAC	cellula Mauthneri	Mauthner cell	230; *58*	15, 98, 99
MaON	nucleus octavus magnocellularis	magnocellular octaval nucleus	237, 239; **15, 22**; *58*	14, 15, 93, 94, 98
MFB	fasciculus medialis telencephali	medial forebrain bundle	98, 107, 114, 121, 125, 127, 131, 136, 149; **34, 53**; *18*	9, 11
MFN	nucleus funiculi medialis	medial funicular nucleus	303, 319; **34**; *46, 58*	14, 16, 94, 95, 97
MLF	fasciculus longitudinalis medialis	medial longitudinal fascicle	173, 179, 185, 196, 201, 204, 208, 213, 219, 223, 230, 237, 239, 251, 260, 272, 279, 283, 290, 303, 319, 363; **46, 53**; *26, 31, 46, 58, 76*	2, 15, 16, 98
MNV	nucleus mesencephalicus nervi trigemini	mesencephalic nucleus of the trigeminal nerve	158	14, 97, 98
MO	medulla oblongata	medulla oblongata		12, 13, 14, 16, 17, 96 97, 99
MON	nucleus octavolateralis medialis	medial octavolateralis nucleus	230, 237, 239, 251, 260; **9, 15, 22, 34**; *31, 46*	14, 92, 93, 99
MOT	tractus olfactorius medialis	medial olfactory tract	50, 60, 71, 85, 92; **34, 46, 53**; *18*	7, 8, 9, 89

Abbr.	Latin	English	Sect. No.	Text (page)
MS	medulla spinalis	spinal cord		1, 3, 4, 7, 12, 15, 16, 17, 94, 96, 97, 98, 99
NC	nucleus commissuralis Cajal	commissural nucleus of Cajal	319; **46**, **53**; *46, 58*	14, 17, 94, 96
NDⅤ	nucleus descendens nervi trigemini	nucleus of the descending trigeminal root	239, 251, 260	13, 97, 98, 99
NI	nucleus isthmi	isthmic nucleus	204, 208, 213; **9**, **15**; *31*	15, 91, 92, 99, 100
NIn	nucleus interpeduncularis	interpeduncular nucleus	185, 196; **46**, **53**; *46, 58*	11, 12, 13, 15, 16
NLL	nucleus lemnisci lateralis	nucleus of the lateral lemniscus (of Prasada Rao et al. 87)	179, 185	12, 99
NLV	nucleus lateralis valvulae	nucleus lateralis valvulae	185, 196, 201, 204, 208; **9**, **15**, **22**, **34**; *18, 26, 31*	15, 16, 91, 99
NMLF	nucleus fasciculi longitudinalis medialis	nucleus of the medial longitudinal fascicle	168, 173, 179; **46**, **53**; *18, 26, 31*	2, 10, 16, 98, 100
NR	nucleus ruber	red nucleus	168	12, 16, 98, 99, 100
NT	nucleus taeniae	nucleus taeniae	98, 107, 114; **9**, **15**; *1*	8, 89
NⅢ	nucleus nervi oculomotorii	oculomotor nucleus	179, 185; **46**, **53**; *31*	12, 13, 97, 100
NⅣ	nucleus nervi trochlearis	trochlear nucleus	196; **53**; *26, 31*	12, 13, 97
NⅤmd	nucleus motorius nervi trigemini, pars dorsalis	trigeminal motor nucleus, dorsal part	208, 213, 219; **22**; *46*	13, 98
NⅤmv	nucleus motorius nervi trigemini, pars ventralis	trigeminal motor nucleus, ventral part	213, 219, 223; **22**; *58*	13, 98
NⅤs	nucleus sensorius principalis nervi trigemini	primary sensory trigeminal nucleus	219	13, 16, 97, 98, 99
NⅥc	nucleus nervi abducentis, pars caudalis	abducens nucleus, caudal part	237, 239	14, 97
NⅦm	nucleus motorius nervi facialis	facial motor nucleus	251; **53**	14, 92, 93, 97, 98
NⅩm	nucleus motorius nervi vagi	vagal motor nucleus	279, 283, 290, 303; **46**, **53**; *58*	14, 96, 98
OB	bulbus olfactorius	olfactory bulb		7, 8, 89
ON	nervus opticus (=Ⅱ)	optic nerve	107; **15**, **22**, **34**, **46**, **53**; *58*	11, 90
OT	tractus opticus	optic tract	107, 114, 121; **34**; *58*	11
P	nucleus posterior thalami	posterior thalamic nucleus	158, 162; *58*	10, 95
PC	tractus cerebellaris posterior	posterior cerebellar tract	219, 223; **22**; *18, 26, 31*	16

Abbr.	Latin	English	Sect. No.	Text (page)
PCN	nucleus paracommissuralis	paracommissural nucleus	149	10, 99
PG	area praeglomerulosa	preglomerular area		11, 90, 95, 96
PGa	nucleus praeglomerulosus anterior	anterior preglomerular nucleus	136; **15, 22**; *76*	10
PGc	nucleus praeglomerulosus caudalis	caudal preglomerular nucleus	173	10
PGl	nucleus praeglomerulosus lateralis	lateral preglomerular nucleus	136, 149, 153, 158; **9, 15, 22**; *58*	10, 92
PGm	nucleus praeglomerulosus medialis	medial preglomerular nucleus	149, 153, 158, 162, 168; **15, 22**; *58*	10
PGZ	stratum periventriculare tecti optici	periventricular gray zone of optic tectum	136, 149, 153, 158, 162, 168, 173, 179, 185, 196, 201, 204, 208, 213, 219, 223; **1, 9, 15, 22, 34, 46, 53**; *1, 18, 26, 31, 46*	12
Pit	hypophysis = glandula pituitaria	hypophysis = pituitary	153, 158, 162, 168, 173	6, 10
PL	nucleus perilemniscularis	perilemniscal nucleus	179, 185, 196; **9**; *31*	12, 99
PLLN	nervus lineae lateralis posterioris	posterior lateral line nerve	251, 260, 272; **15**; *46, 58*	13, 92
PM	nucleus praeopticus magnocellularis	magnocellular preoptic nucleus	114; **46**; *26, 31*	9
PMg	nucleus praeopticus magnocellularis, pars gigantocellularis	gigantocellular part of magnocellular preoptic nucleus	121; **46**	9
PO	nucleus praetectalis posterior	posterior prectectal nucleus (of Wullimann & Meyer 90)	136; **22**; *18, 26, 31*	11
POF	stratum nervosum bulbi olfactorii	primary olfactory fiber layer	23; **15, 22, 34, 46, 53**; *1, 18 , 26*	8
PON	nucleus octavus posterior	posterior octaval nucleus	279; *76*	14, 93, 94
PPa	nucleus praeopticus parvocellularis, pars anterior	parvocellular preoptic nucleus, anterior part	92, 98, 107; **34, 46, 53**; *18, 26, 31, 46*	9
PPd	nucleus praetectalis periventricularis, pars dorsalis	periventricular pretectal nucleus, dorsal part	149, 153; *18*	10, 91, 99, 100
PPp	nucleus praeopticus parvocellularis, pars posterior	parvocellular preoptic nucleus, posterior part	114, 121, 125, 127, 131; **46, 53**; *31, 46*	9
PPv	nucleus praetectalis periventricularis, pars ventralis	periventricular pretectal nucleus, ventral part	149, 153, 158; **34, 46, 53**; *18, 26*	10
PR	recessus posterior	posterior recess of diencephalic ventricle	173; **34, 46, 53**	10

Abbr.	Latin	English	Sect. No.	Text (page)
PSm	nucleus praetectalis super-ficialis, pars magnocellularis	magnocellular superficial pretectal nucleus	127, 131; **9, 15**; *26, 31*	11, 90, 91
PSp	nucleus praetectalis super-ficialis, pars parvocellularis	parvocellular superficial pretectal nucleus	121, 125, 127, 131; **9, 15**; *18, 26, 31*	11, 91
PTN	nucleus tuberis posterior	posterior tuberal nucleus	149, 153, 158, 162; **34, 46, 53**; *58*	10, 11, 89, 101
PVO	organum paraventriculare	paraventricular organ	158	10
R	nucleus rostrolateralis	rostrolateral nucleus (of Butler & Saidel 91)	125	10
RT	nucleus tegmentalis rostralis	rostral tegmental nucleus (of Grover & Sharma 81)	162, 168; **22**	12
RV	ventriculus rhombencephali	rhombencephalic ventricle	201, 204, 208, 213, 219, 223, 230, 237, 239, 251, 260, 272, 279, 283, 290, 303; **46, 53**; *18, 26, 31, 46, 58*	15, 16
SC	nucleus suprachiasmaticus	suprachiasmatic nucleus	114, 121, 125, 127; **53**; *46*	9, 91
SCO	organum subcommissurale	subcommissural organ	149; *1*	10
SD	saccus dorsalis	dorsal sac	121, 125, 127, 131; **46, 53**	9
SG	nucleus subglomerulosus	subglomerular nucleus	168; **22**	10
SGN	nucleus gustatorius secundarius	secondary gustatory nucleus	213; **22**; *26*	13, 15, 16, 95, 96
SGT	tractus gustatorius secundarius	secondary gustatory tract	219, 223, 230, 237, 239, 251, 260, 272, 279, 283, 290, 303; **22**; *58*	14, 16, 95, 96
SO	populatio octavia secundaria	secondary octaval population (of McCormick & Hernandez 96)	237, 239; **34, 53**; *46*	14, 93
SR	nucleus raphes superior	superior raphe	201, 204, 208; **46**; *46, 58*	15, 98, 99
SRF	formatio reticularis, pars superior	superior reticular formation	196, 201, 204, 208, 213, 219, 223; **34, 46, 53**; *46, 58, 76*	12, 14, 15, 16, 98
SY	sulcus ypsiloniformis	ypsiloniform sulcus	85, 92, 98	8
T	nucleus tangentialis	tangential nucleus	237, 239	14, 93, 94, 97, 98
TBS	tractus bulbospinalis	bulbo-spinal tract	260, 272, 279, 283, 290, 303, 319; **22, 34, 46**	17, 98

Abbr.	Latin	English	Sect. No.	Text (page)
Tel	telencephalon	telencephalon		4, 7, 8, 9, 11, 89, 90, 93, 95, 97, 99, 100, 101
TelV	ventriculi telencephali	telencephalic ventricles	50, 60, 85, 98, 107, 114; *1, 18, 26*	8
TeO	tectum opticum	optic tectum	127, 131, 136, 149, 153, 158, 162, 168, 173, 179, 185, 196, 201, 204, 208, 213, 219, 223, 230; **1, 9, 15, 22, 34, 46, 53**; *1, 18, 26, 31, 46, 58*	11, 12, 14, 16, 90, 99 100
TeV	ventriculus mesencephali	tectal ventricle	153, 158, 162, 168, 173, 179, 185, 196; **1, 9, 15, 22, 34, 46, 53**; *1, 18, 26*	12, 13, 14
TGN	nucleus gustatorius tertius	tertiary gustatory nucleus (of Wullimann 88)	162, 168; *58*	10, 91, 95
TH	hypothalamus, pars tuberalis	tuberal hypothalamus		10, 93
TL	torus longitudinalis	longitudinal torus	149, 153, 158, 162, 168, 173, 179, 185, 196, 201, 204; **34, 46, 53**; *1*	12
TLa	torus lateralis	lateral torus	149, 153, 158, 162, 168, 173, 179, 185; **1, 9, 15**; *58, 76*	10
TMCa	tractus mesencephalo-cerebellaris anterior	anterior mesencephalo-cerebellar tract	179, 185, 196; **22**	16
TMCp	tractus mesencephalo-cerebellaris posterior	posterior mesencephalo-cerebellar tract	196; **22**	16
TPM	tractus praetectomamillaris	pretecto-mammillary tract	136, 149, 153, 158, 162, 168, 173, 179; **22**; *31, 46, 58, 76*	11
TPp	nucleus periventricularis tuberculi posterioris	periventricular nucleus of posterior tuberculum	136, 149, 153, 158, 162; **34, 46, 53**; *31, 46*	10
TS	torus semicircularis	semicircular torus	162; **15**	12, 16, 92, 93, 96, 100
TSc	nucleus centralis tori semicircularis	central nucleus of semicircular torus	168, 173, 179, 185, 196, 201, 204, 208 213; **9**; *1, 18,*	12, 93
TSvl	nucleus ventrolateralis tori semicircularis	ventrolateral nucleus of semicircular torus	168, 173, 179, 185, 196, 201, 204, 208, 213; **1, 9**; *18, 26, 31*	12, 92, 100

Abbr.	Latin	English	Sect. No.	Text (page)
TTB	tractus tectobulbaris	tecto-bulbar tract	173, 179, 185, 196, 201, 204, 208; **1, 9, 15, 22**; *31, 46, 58*	16
TTBc	tractus tectobulbaris cruciatus	crossed tecto-bulbar tract	185, 196, 201, 204, 208, 213, 219, 223, 230, 237; **46, 53**; *58, 76*	16
TTBr	tractus tectobulbaris rectus	uncrossed tecto-bulbar tract	196, 201, 204, 208, 213, 219, 223, 230; **15**; *76*	16
TVS	tractus vestibulospinalis	vestibulo-spinal tract	272, 279, 283, 290; **34**	17, 98
V	area ventralis telencephali	ventral telencephalic area	50, 60, 71, 85, 92, 98, 107; **15, 22, 34, 46, 53**; *1, 18, 26, 31*	7, 8, 9, 89, 93, 101
Val	valvula cerebelli, pars lateralis	lateral division of valvula cerebelli	168, 173, 179, 185, 196, 201, 204; **15, 22, 34, 53**; *1*	13, 100
Vam	valvula cerebelli, pars medialis	medial division of valvula cerebelli	179, 185, 196, 201, 204, 208, 213; **34, 46, 53**; *1, 18, 26*	13
VAO	nucleus opticus accessorius ventralis	ventral accessory optic nucleus	136	11, 91, 99
Vas	lacuna vasculosa areae postremae	vascular lacuna of area postrema	168, 173, 179; **46**	
Vc	area ventralis telencephali, nucleus centralis	central nucleus of ventral telencephalic area	60, 71; **53**	8, 89
Vd	area ventralis telencephali, nucleus dorsalis	dorsal nucleus of ventral telencephalic area	50, 60, 71, 85; **34, 46**; *1, 18*	8, 89, 101
VH	cornu ventrale	ventral horn	363	15
Vl	area ventralis telencephali, nucleus lateralis	lateral nucleus of ventral telencephalic area	60, 71; **15, 22, 53**; *18, 26, 31*	8
VL	nucleus ventrolateralis thalami	ventrolateral thalamic nucleus	121, 125, 127, 131, 136; *18, 26*	10, 90, 96, 97, 100
VM	nucleus ventromedialis thalami	ventromedial thalamic nucleus	114, 121, 125, 127, 131, 136; *18, 26*	10, 93, 96, 97, 100
VOT	tractus opticus ventrolateralis	ventrolateral optic tract	125, 127, 131, 136, 149, 153, 158, 162; **1, 9, 15, 22**; *46*	11
Vp	area ventralis telencephali, nucleus postcommissuralis	postcommissural nucleus of ventral telencephalic area	98, 107; **34, 46**; *1*	8

Abbr.	Latin	English	Sect. No.	Text (page)
Vs	area ventralis telencephali, nucleus supracommissuralis	supracommissural nucleus of ventral telencephalic area	92; **34**, **46**, **53**; *1*	8, 9
VT	thalamus, pars ventralis	ventral thalamus	**34**, **46**, **53**	4, 9, 10, 97
Vv	area ventralis telencephali, nucleus ventralis	ventral nucleus of ventral telencephalic area	60, 71, 85; **34**, **46**, **53**; *18, 26, 31*	8, 9, 89
ZL	zona limitans	zona limitans	136; **46**	
I	nervus olfactorius	olfactory nerve		7, 89
II	nervus opticus	optic nerve (=ON)	107; **15**, **22**, **34**, **46**, **53**; *58*	11, 90
III	nervus oculomotorius	oculomotor nerve	179; **1**, **9**, **15**, **22**, **34**, **46**, **53**; *46, 58*	12
IV	nervus trochlearis	trochlear nerve	201, 208, 213; **1**, **9**, **15**, **22**; *26, 31*	12, 92
V	nervus trigeminus	trigeminal nerve	213; **1**; *58*	13, 14, 95, 97
Vmd	radix motoria nervi trigemini, pars dorsalis	dorsal motor root of the trigeminal nerve	219	
Vmv	radix motoria nervi trigemini, pars ventralis	ventral motor root of the trigeminal nerve	223	
Vs	radix sensoria nervi trigemini	sensory root of the trigeminal nerve	219; **9**, **15**; *46*	16, 97
VI	nervus abducens	abducens nerve		13, 14
VIc	radix caudalis nervi abducentis	caudal root of the abducens nerve	237	14
VIr	radix rostralis nervi abducentis	rostral root of the abducens nerve	223	14
VII	nervus facialis	facial nerve	219, 223; **1**; *58*	13, 92, 94
VIIs	radix sensoria nervi facialis	sensory root of the facial nerve	230, 237, 239, 251, 260; **9**, **15**, **22**, **34**, **53**; *46*	14, 95
VIII	nervus octavus	octaval nerve	230, 237, 239, 251, 260, 272; **9**, **15**; *58, 76*	13, 14, 93, 94
IX	nervus glossopharyngeus	glossopharyngeal nerve	272	13, 14, 94
X	nervus vagus	vagal nerve	279, 283, 290; *46, 58*	13, 14, 92, 94, 95, 96

10 References

Akimenko, M.A., Ekker, M., Wegner, J., Lin, W., Westerfield, M. (1994) Combinatorial expression of three zebrafish genes related to *Distal-less*: part of a homeobox gene code for the head. *J. Neurosci.* 14: 3475-3486.

Allum, J.H.J., Greef, N.G., Tokunaga, A. (1981) Projections of the rostral and caudal abducens nuclei in the goldfish. *In:* Fuchs A.F., Becker W. (eds.): *Progress in Oculomotor Research.* Amsterdam: Elsevier, pp. 253-262.

Anken, R.H., Rahmann, H. (1994) *Brain Atlas of the Adult Swordtail Fish Xiphophorus helleri and of Certain Developmental Stages.* Stuttgart: G. Fischer Verlag.

Anken, R.H., Rahmann, H. (1995) Notes on the organization of the rostral diencephalon of the atherinomorph swordtail fish *Xiphophorus helleri. Ann. Anat.* 177: 51-59.

Baier, H., Korsching, S. (1994) Olfactory glomeruli in the zebrafish form an invariant pattern and are identifiable across animals. *J. Neurosci.* 14: 219-230.

Barman, R.P. (1991) A taxonomic revision of the Indo-Burmese species of *Danio* Hamilton-Buchanan (Pisces, Cyprinidae). *Rec. Zool. Surv. India Occas. Pap.* Vol. 137.

Bartheld, C.S. von (1987) Central connections of the terminal nerve in ray-finned fishes. *Ann. N. Y. Acad. Sci.* 519: 392-410.

Bartheld, C.S. von, Meyer, D.L. (1987) Comparative neurology of the optic tectum in ray-finned fishes: patterns of lamination formed by retinotectal projections. *Brain Res.* 420: 277-288.

Bartheld, C.S. von, Meyer, D.L., Fiebig, E., Ebbesson, S. (1984) Central connections of the olfactory bulb in the goldfish, *Carassius auratus. Cell Tiss. Res.* 238: 475-487.

Becker, T., Becker, C.G., Bernhardt, R.R., Tongiorgi, E., Wullimann, M.F., Schachner, M (1995) Brain nuclei with axons projecting to the spinal cord in zebrafish: regenerative capacity and regulation of cell adhesion molecules. *Soc. Neurosci. Abstr.* 21:1279.

Bernhardt, R.R., Chitnis, A.B., Lindamer, L., Kuwada, J.Y. (1990) Identification of spinal neurons in the embryonic and larval zebrafish. *J. Comp. Neurol.* 302: 603-616.

Bleckmann, H., Niemann, U., Fritzsch, B. (1991) Peripheral and central aspects of the acoustic and lateral line system of a bottom dwelling catfish *Ancistrus sp.. J. Comp. Neurol.* 314: 452-466.

Bosch, T.J., Paul, D.H. (1993) Differential responses of single reticulospinal cells to spatially localized stimulation of the optic tectum in a teleost fish, *Salmo trutta. Europ. J. Neurosci.* 5: 742-750.

Braford, M.R. jr., Northcutt, R.G. (1983) Organization of the diencephalon and pretectum of the ray-finned fishes. *In:* Davis R.E., Northcutt R.G. (eds.): *Fish Neurobiology,* Vol. 2 *Higher Brain Areas and Functions.* Ann Arbor: University of Michigan Press, pp. 117-140.

Bulfone, A., Puelles, L., Porteus, M.H., Frohman, M.A., Martin, G.R., Rubenstein, J.L.R. (1993) Spatially restricted expression of *Dlx-1, Dlx-2 (Tes-1), Gbx-2,* and *Wnt-3* in the embryonic day 12.5 mouse forebrain defines potential transverse and longitudinal segmental boundaries. *J. Neurosci.* 13: 3155-3172.

Burrill, J.D., Easter, S.S. jr. (1994) Development of the retinofugal projections in the embryonic and larval zebrafish (*Brachydanio rerio*). *J. Comp. Neurol.* 346: 583-600.

Butler, A.B., Saidel, W.M. (1991) Retinal projections in the freshwater butterfly fish, *Pantodon buchholzi* (Osteoglossoidei). I. Cytoarchitectonic analysis and primary visual pathways. *Brain Behav. Evol.* 38: 127-153.

Butler, A.B., Wullimann, M.F., Northcutt, R.G. (1991) Comparative cytoarchitectonic analysis of some visual pretectal nuclei in teleosts. *Brain Behav. Evol.* 38: 92-114.

Chitnis, A.B., Kuwada, J.Y. (1990) Axonogenesis in the brain of zebrafish embryos. *J. Neurosci.* 10: 1892-1905.

Chitnis, A.B., Kuwada, J.Y. (1991) Elimination of a brain tract increases errors in pathfinding by follower growth cones in the zebrafish embryo. *Neuron* 7: 277-285.

Chitnis, A.B., Patel, C.K., Kim, S., Kuwada, J.Y. (1992) A specific brain tract guides follower growth cones in two regions of the zebrafish brain. *J. Neurobiol.* 23: 845-854.

Clarke, J.D.W., Lumsden, A. (1993) Segmental repetition of neuronal phenotypes in the chick embryo. *Development* 118: 151-162.

Coombs, S., Görner, P., Münz, H. 1989, *The Mechanosensory Lateral Line.* New York: Springer Verlag.

Demski, L.S., Dulka, J.G. (1984) Functional-anatomical studies on sperm-release evoked by electrical stimulation of the olfactory tract in goldfish. *Brain Res.* 291: 241-247.

Demski, L.S., Sloan, H.E. (1985) A direct magnocellular-preopticospinal pathway in goldfish: implications for control of sex behavior. *Neurosci. Lett.* 55: 283-288.

Douglas, R.H., Djamgoz, M.B.A. (1990) *The Visual System of Fish.* London: Chapman & Hall.

Douglas, R.H., Hawryshyn, C.W. (1990) Behavioural studies of fish vision: an analysis of visual capabilities. *In:* Douglas R.H., Djamgoz M.B.A. (eds.): *The Visual System of Fish.* London: Chapman & Hall, pp. 373-418.

Dulka, J.G., Stacey, N.E., Sörensen, P.W., Van der Kraak, G.J. (1987) A steroid sex pheromone synchronizes male-female spawning readiness in goldfish. *Nature* 325: 251-253.

Ebbesson, S.O.E., Schroeder, D.M. (1971) Connections of the nurse shark's telencephalon. *Science* 173: 254-256.

Echteler, S.M. (1984) Connections of the auditory midbrain in a teleost fish, *Cyprinus carpio. J. Comp. Neurol.* 230: 536-551.

Echteler, S.M. (1985) Organization of central auditory pathways in a teleost fish, *Cyprinus carpio. J. Comp. Physiol. A* 156: 267-280.

Echteler, S.M., Saidel, W.M. (1981) Forebrain connections in the goldfish support telencephalic homologies with land vertebrates. *Science.* 212: 683-685.

Eisen, J.S., Myers, P.Z., Westerfield, M. (1986) Pathway selection by growth cones of identified motoneurones in live zebrafish embryos. *Nature* 320: 269-271.

Ekker, M., Wegner, J., Akimenko, M.A., Westerfield, M. (1992) Coordinate embryonic expression of three zebrafish *engrailed* genes. *Development* 116: 1001-1010.

Ekström, P., Honkanen, T., Steinbusch, H.W.M. (1990) Distribution of dopamine-immunoreactive neuronal perikarya and fibers in the brain of a teleost, *Gasterosteus aculeatus* L. Comparison with TH- and DBH-IR neurons. *J. Chem. Neuroanat.* 3: 233-260.

Ekström, P., Reschke, M., Steinbusch, H.W.M., Veen, T. van (1986) Distribution of noradrenaline in the brain of the teleost *Gasterosteus aculeatus* L.: an immunohistochemical analysis. *J. Comp. Neurol.* 254: 297-313.

Fay, R.R. (1992) Analytic listening by the goldfish. *Hearing Res.* 59: 101-107.

Figdor, M.C., Stern, C.D. (1993) Segmental organization of embryonic diencephalon. *Nature* 363: 630-634.

Finger, T.E. (1978a) Gustatory pathways in the bullhead catfish. II. Facial lobe connections. *J. Comp. Neurol.* 180: 691-706.

Finger, T.E. (1978b) Cerebellar afferents in teleost catfish (Ictaluridae). *J. Comp. Neurol.* 181: 173-181.

Finger, T.E. (1981) Fish that taste with their feet: spinal sensory pathways in the sea robin, *Prionotus carolinus. Biol. Bull.* 161: 343.

Finger, T.E. (1983) Organization of the teleost cerebellum. *In:* Northcutt R.G., Davis R.E. (eds.): *Fish Neurobiology, Vol. 1 Brain Stem and Sense Organs.* Ann Arbor: University of Michigan Press, pp. 261-284.

Finger, T.E. (1986) Electroreception in catfish:, behavior, anatomy and electrophysiology. *In:* Bullock T.H., Heiligenberg W. (eds.): *Electroreception.* New York: Wiley, pp. 287-317.

Finger, T.E., Kanwal, J.S. (1992) Ascending general visceral pathways within the brainstem of two teleost fishes: *Ictalurus punctatus* and *Carassius auratus. J. Comp. Neurol.* 320: 509-520.

Finger, T.E., Karten, H.J. (1978) The accessory optic system in teleosts. *Brain Res.* 153: 144-149.

Fjose, A., Izpisua-Belmonte, J.C., Fromental-Ramain, C., Duboule, D. (1994) Expression of the zebrafish gene *hlx-1* in the prechordal plate and during CNS development. *Development* 120: 71-81.

Fjose, A., Njølstad, P.R., Nornes, S., Molven, A., Krauss, S. (1992) Structure and early embryonic expression of the zebrafish *engrailed-2* gene. *Mech. Dev.* 39: 51-62.

Goehler, L., Finger, T. (1992) Functional organization of vagal reflex systems in the brainstem of the goldfish, *Carassius auratus. J. Comp. Neurol.* 319: 463-478.

Graf, W., McGurk, J.F. (1985) Peripheral and central oculomotor organization in the goldfish, *Carassius auratus. J. Comp. Neurol.* 239: 391-401.

Grover, B.G., Sharma, S.C. (1979) Tectal projections in the goldfish (*Carassius auratus*): a degeneration study. *J. Comp. Neurol.* 184: 435-454.

Grover, B.G., Sharma, S.C. (1981) Organization of extrinsic tectal connections in goldfish (*Carassius auratus*). *J. Comp. Neurol.* 196: 471-488.

Guthrie, D.M. (1990) The physiology of the teleostean optic tectum. *In:* Douglas R.H., Djamgoz M.B.A. (eds.): *The Visual System of Fish.* London: Chapman & Hall, pp. 279-343.

Hanneman, E., Westerfield, M. (1989) Early expression of acetylcholinesterase activity in functionaly distinct neurons of the zebrafish. *J. Comp. Neurol.* 284: 350-361.

Hanneman, E., Trevarrow, B., Metcalfe, W., Kimmel, C.B., Westerfield, M. (1988) Segmental development of the spinal cord and hindbrain of the zebrafish embryo. *Development* 103: 49-58.

Hansen, A., Zeiske, E. (1993) Development of the olfactory organ in the zebrafish, *Brachydanio rerio. J. Comp. Neurol.* 333: 289-300.

Hara, T.J. (1992) *Fish Chemoreception.* London: Chapman & Hall.

Hatta, K., Bremiller, R.A., Westerfield, M., Kimmel, C.B. (1991) Diversity of expression of *engrailed*-like antigens in zebrafish. *Development* 112: 821-832.

Hayama, T., Caprio, J. (1989) Lobule structure and somatotopic organization of the medullary facial lobe in the channel catfish *Ictalurus punctatus. J. Comp. Neurol.* 285: 9-17.

Herrick, C.J. (1905) The central gustatory paths in the brains of bony fishes. *J. Comp. Neurol.* 15: 375-456.

Herrick, C.J. (1948) *The Brain of the Tiger Salamander Ambystoma tigrinum.* Chicago: University of Chicago Press.

Highstein, S.M., Kitch, R., Carey, J., Baker, R. (1992) Anatomical organization of the brainstem octavolateralis area of the oyster toadfish, *Opsanus tau. J. Comp. Neurol.* 319: 501-518.

Holland, P.W.H., Hogan, B.L.M. (1988) Expression of homeobox genes during mouse development: a review. *Genes & Development* 2: 773-782.

Ito, H., Yoshimoto, M. (1990) Cytoarchitecture and fiber connections of the nucleus lateralis valvulae in the carp (*Cyprinus carpio*). *J. Comp. Neurol.* 298: 385-399.

Ito, H., Murakami, T., Fukuoka, T., Kishida, R. (1986) Thalamic fiber connections in a teleost (*Sebastiscus marmoratus*): visual, somatosensory, octavolateral, and cerebellar relay region to the telencephalon. *J. Comp. Neurol.* 250: 215-227.

Kah, O., Chambolle, P. (1983) Serotonin in the brain of the goldfish, *Carassius auratus.* An immunocytochemical study. *Cell Tiss. Res.* 234: 319-333.

Kahn, P. (1994) Zebrafish hit the big time. *Science* 264: 904-905.

Kanwal, J.S., Caprio, J. (1987) Central projections of the glossopharyngeal and vagal nerves in the channel catfish, *Ictalurus punctatus:* clues to differential processing of visceral inputs. *J. Comp. Neurol.* 264: 216-230.

Kanwal, J.S., Caprio, J. (1988) Overlapping taste and tactile maps of the oropharynx in the vagal lobe of the channel catfish, *Ictalurus punctatus:, J. Neurobiol.* 19: 211-222.

Kanwal, J.S., Finger, T.E., Caprio, J. (1988) Forebrain connections of the gustatory system in ictalurid fishes. *J. Comp. Neurol.* 278: 353-376.

Keynes, R.J., Stern, C.D. (1988) Mechanisms of vertebrate segmentation.., *Development* 103: 413-429.

Kimmel, C.B. (1993) Patterning the brain of the zebrafish embryo. *Annu. Rev. Neurosci.* 16: 707-732.

Kimmel, C.B., Metcalfe, W.K., Schabtach, E. (1985) T reticular interneurons: a class of serially repeating cells in the zebrafish hindbrain. *J. Comp. Neurol.* 233: 365-376.

King, W.M., Schmidt, J.T. (1991) The long latency component of retinotectal transmission: enhancement by stimulation of nucleus isthmi or tectobulbar tract and block by nicotinic cholinergic antagonists. *Neurosci.* 40: 701-712.

King, W.M., Schmidt, J.T. (1993) Nucleus isthmi in goldfish: *in vitro* recordings and fiber connections revealed by HRP injections. *Vis. Neurosci.* 10: 419-437.

Kiyohara, S., Toshihiro, S., Yamashita, S. (1985) Peripheral and central distribution of major branches of the facial taste nerve in the carp. *Brain Res.* 325: 57-69.

Kiyohara, S., Hidaka, J., Kitoh, J., Yamashita, S. (1985) Mechanical sensitivity of the facial nerve fibers innervating the anterior palate of the puffer, *Fugu paradalis*, and their central projection to the primary taste center. *J. Comp. Physiol. A* 157: 705-716.

Kiyohara, S., Houman, H., Yamashita, S., Caprio, J., Marui, T. (1986) Morphological evidence for a direct projection of trigeminal nerve fibers to the primary gustatory center in the sea catfish *Plotosus anguillaris*. *Brain Res.* 379: 353-357.

Knudsen, E.I. (1977) Distinct auditory and lateral line nuclei in the midbrain of catfishes. *J. Comp. Neurol.* 173: 417-432.

Krauss, S., Johansen, T., Korzh, V., Fjose, A. (1991a) Expression of the zebrafish paired box gene *pax[zf-b]* during early neurogenesis. *Development* 113: 1193-1206.

Krauss, S., Johansen, T., Korzh, V., Fjose, A. (1991b) Expression pattern of zebrafish *pax* genes suggests a role in early brain regionalization. *Nature* 353: 267-270.

Krauss, S., Korzh, V., Fjose, A., Johansen, T. (1992a) Expression of four zebrafish *wnt*-related genes during embryogenesis. *Development* 116: 249-259.

Krauss, S., Maden, M., Holder, N., Wilson, S.W. (1992b) Zebrafish *pax[b]* is involved in the formation of the midbrain-hindbrain boundary. *Nature* 360: 87-89.

Kuwada, J.Y., Bernhardt, R.R., Nguyen, N. (1990) Development of spinal neurons and tracts in the zebrafish embryo. *J. Comp. Neurol.* 302: 617-628.

Lamb, C.F., Caprio, J. (1993) Diencephalic gustatory connections in the channel catfish. *J. Comp. Neurol.* 337: 400-418.

Lauder, G.V., Liem, K.F. (1983) The evolution and interrelationships of the actinopterygian fishes. *Bull. Mus. Comp. Zool.* 150: 95-197.

Lee, R.K.K., Bullock T. (1984) Sensory representation in the cerebellum of the catfish. *Neurosci.* 13: 157-169.

Lee, R.K.K., Eaton, R.C. (1991) Identifiable reticulospinal neurons of the adult zebrafish, *Brachydanio rerio. J. Comp. Neurol.* 304: 34-52.

Lee, R.K.K., Eaton, R.C., Zottoli, S.J. (1993a) Segmental arrangement of reticulospinal neurons in the goldfish hindbrain. *J. Comp. Neurol.* 329: 539-556.

Lee, R.K.K., Finger, T.E., Eaton, R.C. (1993b) GABAergic innervation of the Mauthner cell and other reticulospinal neurons in the goldfish. *J. Comp. Neurol.* 338: 601-611.

Leiner, H.C., Leiner, A.L., Dow, R.S. (1991) The human cerebro-cerebellar system: its computing, cognitive, and language skills. *Behav. Brain Res.* 44: 113-128.

Levine, R.L., Dethier, S. (1985) The connections between the olfactory bulb and the brain in the goldfish. *J. Comp. Neurol.* 237: 427-444.

Luiten, P.G.M. (1975) The central projections of the trigeminal, facial and anterior lateral line nerves in the carp (*Cyprinus carpio* L.). *J. Comp. Neurol.* 160: 399-417.

Luiten, P.G.M. (1976) A somatotopic and functional representation of the respiratory muscles in the trigeminal and facial motor nuclei of the carp (*Cyprinus carpio* L.). *J. Comp. Neurol.* 166: 191-200.

Luiten, P.G.M. (1979) Proprioceptive reflex connection of head musculature and the mesencephalic trigeminal nucleus in the carp. *J. Comp. Neurol.* 183: 903-912.

Luiten, P.G.M. (1981) Afferent and efferent connections of the optic tectum in the carp (*Cyprinus carpio* L.). *Brain Res.* 220: 51-65.

Luiten, P.G.M., Dijkstra-de Vlieger, H.P. (1978) Extraocular muscle representation in the brainstem of the carp. *J. Comp. Neurol.* 179: 669-676.

Luiten, P.G.M., van der Pers, J.N.C. (1977) The connections of the trigeminal and facial motor nuclei in the brain of the carp (*Cyprinus carpio* L.) as revealed by anterograde and retrograde transport of HRP. *J. Comp. Neurol.* 174: 575-590.

Lumsden, A. (1990) The cellular basis of segmentation in the developing hindbrain. *TINS* 13: 329-339.

Lumsden, A., Keynes, R. (1989) Segmental patterns of neuronal development in the chick hindbrain. *Nature* 337: 424-428.

Ma, P.M. (1994a) Catecholaminergic systems in the zebrafish. I. Number, morphology, and histochemical characteristics of neurons in the locus coeruleus. *J. Comp. Neurol.* 344: 242-255.

Ma, P.M. (1994b) Catecholaminergic systems in the zebrafish. II. Projection pathways and pattern of termination of the locus coeruleus. *J. Comp. Neurol.* 344: 256-269.

Macdonald, R., Xu, Q., Barth, K.A., Mikkola, I., Holder, N., Fjose, A., Krauss, S., Wilson, S.W. (1994) Regulatory gene expression boundaries demarcate sites of neuronal differentiation in the embryonic zebrafish forebrain. *Neuron* 13: 1039-1053.

Marin, F., Puelles, L. (1994) Patterning of the embryonic avian midbrain after experimental inversions: a polarizing activity from the isthmus. *Dev. Biol.* 163: 19-37.

Marui, T., Caprio, J., Kiyohara, S., Kasahara, Y. (1988) Topographical organization of taste and tactile neurons in the facial lobe of the sea catfish *Plotosus lineatus*. *Brain Res.* 446: 178-182.

McCormick, C.A. (1982) The organization of the octavolateralis area in actinopterygian fishes: a new interpretation. *J. Morphol.* 171: 159-181.

McCormick, C.A. (1989) Central lateral line mechanosensory pathways in bony fish. *In:* Coombs S., Görner P., Münz P (eds.): *The Mechanosensory Lateral Line.* New York: Springer Verlag, pp. 341-364.

McCormick, C.A. (1992) Evolution of central auditory pathways in anamniotes. *In:* Webster D.B., Fay R.R., Popper A.N. (eds.): *The Evolutionary Biology of Hearing.* New York: Springer Verlag, pp. 323-350.

McCormick, C.A., Braford, M.R. jr. (1994) Organization of inner ear endorgan projections in the goldfish, *Carassius auratus*. *Brain Behav. Evol.* 43: 189-205.

McCormick, C.A., Hernandez, D.V. (1996) Connections of the octaval and lateral line nuclei of the medulla in the goldfish, including the cytoarchitecture of the secondary octaval population in goldfish and catfish. *Brain Behav. Evol.*: 47 *(in press)*.

Medina, M., Reperant, J., Ward, R., Rio, J.P., Lemire, M. (1993) The primary visual system of flatfish: an evolutionary perspective. *Anat. Embryol.* 187: 167-191.

Meek, J. (1983) Functional anatomy of the tectum mesencephali of the goldfish: an explorative analysis of the functional implications of the laminar structural organization of the tectum. *Brain Res. Rev.* 6: 247-297.

Meek, J. (1990) Tectal morphology: connections, neurons and synapses. *In:* Douglas R.H., Djamgoz M.B.A. (eds.): *The Visual System of Fish.* London: Chapman & Hall, pp. 239-277.

Meek, J. (1992) Why run parallel fibers parallel? Teleostean Purkinje cells as possible coincidence detectors in a timing device subserving spatial coding of temporal differences. *Neurosci.* 48: 249-283.

Mendelson, B. (1986) Development of reticulospinal neurons of the zebrafish. II. Early axonal outgrowth and cell body position. *J. Comp. Neurol.* 251: 172-184.

Meredith, G.E. (1985) The distinctive central utricular projections in the herring. *Neurosci. Lett.* 55: 191-196.

Meredith, G.E., Roberts, B.L., Maslam, S. (1987) Distribution of afferent fibers in the brainstem from end organs in the ear and lateral line in the European eel. *J. Comp. Neurol.* 265: 507-520.

Metcalfe, W.K., Kimmel, C.B., Schabtach, E. (1985) Anatomy of the posterior lateral line system in young larvae of the zebrafish. *J. Comp. Neurol.* 233: 377-389.

Metcalfe, W.K., Mendelson, B., Kimmel, C.B. (1986) Segmental homologies among reticulospinal neurons in the hindbrain of the zebrafish larva. *J. Comp. Neurol.* 251: 147-159.

Metcalfe, W.K., Myers, P.Z., Trevarrow, B., Bass, M.B., Kimmel, C.B. (1990) Primary neurons that express the L2/HNK-1 carbohydrate during early development in the zebrafish. *Development* 110: 491-504.

Mikkola, I., Fjose, A., Kuwada, J.Y., Wilson, S.W., Guddal, P.H., Krauss, S. (1992) The paired domain-containing nuclear factor *pax[b]* is expressed in specific commissural interneurons in zebrafish embryos. *J. Neurobiol.* 23: 933-946.

Molven, A., Njølstad, P.R., Fjose, A. (1991) Genomic structure and restricted neural expression of the zebrafish *wnt-1 (int-1)* gene. *EMBO J.* 10: 799-807.

Mori, S. (1993) Localization of extratectally evoked visual response in the corpus and valvula cerebelli in carp, and cerebellar contribution to „dorsal light reaction" behavior. *Behav. Brain Res.* 59: 33-40.

Morita, Y., Finger, T.E. (1985) Topographic and laminar organization of the vagal gustatory system in the goldfish, *Carassius auratus. J. Comp. Neurol.* 238: 187-201.

Morita, Y., Finger, T.E. (1987) Topographic representation of the sensory and motor roots of the vagus nerve in the medulla of goldfish, *Carassius auratus. J. Comp. Neurol.* 264: 231-249.

Morita, Y., Ito, H., Masai, H. (1980) Central gustatory paths in the crucian carp, *Carassius carassius. J. Comp. Neurol.* 191: 119-132.

Morita, Y., Murakami, T., Ito, H. (1983) Cytoarchitecture and topographic projections of the gustatory centers in a teleost, *Carassius carassius. J. Comp. Neurol.* 218: 378-394.

Mullins, M.C., Nüsslein-Volhard, C. (1993) Mutational approaches to study embryonic pattern formation in the zebrafish. *Curr. Opin. Genet. Dev.* 3: 648-654.

Munoz, A., Munoz, M., Gonzalez, A., Ten Donkelaar, H.J. (1994) Spinothalamic projections in amphibians as revealed with anterograde tracing techniques. *Neurosci. Lett.* 171: 81-84.

Murakami, T., Ito, H. (1985) Long ascending projections of the spinal dorsal horn in a teleost, *Sebastiscus marmoratus. Brain Res.* 346: 168-170.

Murakami, T., Fukuoka, T., Ito, H. (1986a) Telencephalic ascending acousticolateral system in a teleost, *Sebastiscus marmoratus*, with special reference to fiber connections of the nucleus preglomerulosus. *J. Comp. Neurol.* 247: 383-397.

Murakami, T., Ito, H., Morita, Y. (1986b) Telencephalic afferent nuclei in the carp diencephalon, with special reference to fiber connections of the nucleus praeglomerulosus pars lateralis. *Brain Res.* 382: 97-103.

Murakami, T., Morita, Y., Ito, H. (1983) Extrinsic and intrinsic fiber connections of the telencephalon in a teleost, *Sebastiscus marmoratus. J. Comp. Neurol.* 216: 115-131.

Murphy, P., Davidson, D.R., Hill, R.E. (1989) Segment-specific expression of a homeobox-containing gene in the mouse hindbrain. *Nature* 341: 156-159.

Myers, P.Z. (1985) Spinal motoneurons of the larval zebrafish. *J. Comp. Neurol.* 236: 555-561.

Neumeyer, C. (1992) Tetrachromatic color vision in goldfish: evidence from color mixture experiments. *J. Comp. Physiol. A* 171: 639-649.

Nieuwenhuys, R. (1959) The structure of the telencephalon of the teleost *Gasterosteus aculeatus. K. Ned. Akad. Wet. Proc. ser. C.* 62: 341-362.

Nieuwenhuys, R. (1963) The comparative anatomy of the actinopterygian forebrain. *J. Hirnforsch.* 6: 171-192.

Nieuwenhuys, R. (1967) Comparative anatomy of the cerebellum. *Progr. Brain Res.* 25: 1-93.

Nieuwenhuys, R., Meek, J. (1990) The telencephalon of actinopterygian fishes. *In:* Jones E.G., Peters A. (eds.): *Cerebral Cortex*, Vol. 8A. New York: Plenum Press, pp. 31-73.

Nieuwenhuys, R., Pouwels, E. (1983) The brain stem of actinopterygian fishes. *In:* Northcutt R.G., Davis R.E. (eds.): *Fish Neurobiology*, Vol. 1 *Brain Stem and Sense Organs.* Ann Arbor: University of Michigan Press, pp. 25-87.

Nieuwenhuys, R., Voogd, J., Van Huijzen, C. (1988) *The Human Central Nervous System: A Synopsis and Atlas*. 3rd edition, Berlin: Springer Verlag.

Niida, A., Ohono, T., Iwata, K.S. (1989) Efferent tectal cells of crucian carp: physiology and morphology. *Brain Res. Bull.* 22: 389-398.

Njølstad, P.R., Fjose, A. (1988) In situ hybridization patterns of zebrafish homeobox genes homologous to *Hox-2.1* and *En-2* of mouse. *Biochem. Biophys. Res. Commun.* 157: 426-432.

Njølstad, P.R., Molven, A., Apold, J., Fjose, A. (1990) The zebrafish homeobox gene *hox-2.2*: transcription unit, potential regulatory regions and *in situ* localization of transcripts. *EMBO J.* 9: 515-524.

Northcutt, R.G. (1981a) Localization of neurons afferent to the telencephalon in a primitive bony fish, *Polypterus palmas. Neurosci. Lett.* 22: 219-222.

Northcutt, R.G. (1981b) Evolution of the telencephalon in non-mammals. *Ann. Rev. Neurosci.* 4: 301-350.

Northcutt, R.G. (1983) Evolution of the optic tectum in ray-finned fishes. *In:* Davis R.E., Northcutt R.G. (eds.): *Fish Neurobiology*, Vol. 2 *Higher Brain Areas and Functions.* Ann Arbor: University of Michigan Press, pp. 1-42.

Northcutt, R.G. (1989) The phylogenetic distribution and innervation of craniate mechanoreceptive lateral lines. *In:* Coombs S., Görner P., Münz P. (eds.): *The Mechanosensory Lateral Line.* New York: Springer Verlag, pp. 17-78.

Northcutt, R.G., Braford, M.R. jr. (1980) New observations on the organization and evolution of the telencephalon of actinopterygian fishes. *In:* Ebbesson S.O.E. (ed.): *Comparative Neurology of the Telencephalon.* New York: Plenum Press, pp. 41-98.

Northcutt, R.G., Braford, M.R. jr. (1984) Some efferent connections of the superficial pretectum in the goldfish. *Brain Res.* 296: 181-184.

Northcutt, R.G., Butler, A.B. (1993) The diencephalon of the Pacific herring, *Clupea harengus*: retinofugal projections to the diencephalon and optic tectum. *J. Comp. Neurol.* 328: 547-561.

Northcutt, R.G., Davis, R.E. (1983) Telencephalic organization in ray-finned fishes. *In:* Davis R.E., Northcutt R.G. (eds.): *Fish Neurobiology*, Vol. 2 *Higner brain areas and Functions.* Ann Arbor: University of Michigan Press, pp. 203-236.

Northcutt, R.G., Wullimann, M.F. (1988) The visual system in teleost fishes: morphological patterns and trends. *In:* Atema J., Fay R.R., Popper A.N., Tavolga W.N. (eds.): *Sensory Biology of Aquatic Animals.* New York: Springer Verlag, pp. 515-552.

Northmore, D.P.M. (1991) Visual responses of nucleus isthmi in a teleost fish (*Lepomis macrochirus*). *Vision Res.* 31: 525-535.

Oka, Y., Satou, M., Ueda, K. (1986a) Descending pathways in the himé salmon (landlocked red salmon, *Oncorhynchus nerka*). *J. Comp. Neurol.* 254: 91-103.

Oka, Y., Satou, M., Ueda, K. (1986b) Ascending pathways from the spinal cord in the himé salmon (landlocked red salmon, *Oncorhynchus nerka*). *J. Comp. Neurol.* 254: 104-112.

Oxtoby, E., Jowett, T. (1993) Cloning of the zebrafish *krox-20* gene (*krx-20*) and its expression during hindbrain development. *Nucleic Acids Research* 21: 1087-1095.

Popper, A.N., Fay, R.R. (1993) Sound detection and processing by fish: critical review and major research questions. *Brain Behav. Evol.* 41: 14-38.

Postlethwait, J.H., Johnson, S.L., Midson, C.N., Talbot, W.S., Gates, M., Ballinger, E.W., Africa, D., Andrews, R., Carl, T., Eisen, J.S., Horne, S., Kimmel, C.B., Hutchinson, M., Johnson, M., Rodriguez, A. (1994) A genetic linkage map for the zebrafish. *Science* 264: 699-703.

Prasada Rao, P.D., Jadhao, A.G., Sharma, S.C. (1987) Descending projection neurons to the spinal cord of the goldfish, *Carassius auratus. J. Comp. Neurol.* 265: 96-108.

Puelles, L., Rubenstein, J.L.R (1993) Expression patterns of homeobox and other putative regulatory genes in the embryonic mouse forebrain suggest a neuromeric organization. *TINS* 16: 472-479.

Püschel, A.W., Gruss, P., Westerfield, M. (1992a) Sequence and expression pattern of *pax-6* are highly conserved between zebrafish and mice. *Development* 114: 643-651.

Püschel, A.W., Westerfield, M., Dressler, G.R. (1992b) Comparative analysis of *pax-2* protein distributions during neurulation in mice and zebrafish. *Mech. Dev.* 38: 197-208.

Puzdrowski, R.L. (1987) The peripheral distribution and central projections of the sensory rami of the facial sensory nerve in goldfish, *Carassius auratus. J. Comp. Neurol.* 259: 382-392.

Puzdrowski, R.L. (1988) Afferent projections of the trigeminal nerve in the goldfish, *Carassius auratus. J. Morphol.* 198: 131-147.

Puzdrowski, R.L. (1989) Peripheral distribution and central projections of the lateral line nerves in goldfish, *Carassius auratus. Brain Behav. Evol.* 34: 110-131.

Reiner, A.J., Northcutt, R.G. (1992) An immunohistochemical study of the telencephalon of the Senegal bichir (*Polypterus senegalus*). *J. Comp. Neurol.* 319: 359-386.

Riddle, D.R., Oakley, B. (1991) Evaluation of projection patterns in the primary olfactory system of rainbow trout. *J. Neurosci.* 11: 3752-3762.

Roberts, B.L., Meredith, G.E. (1989) The efferent system. *In:* Coombs S., Görner P., Münz P. (eds): *The Mechanosensory Lateral Line.* New York: Springer Verlag, pp. 445-460.

Roberts, B.L., Meredith, G.E. (1992) The efferent innervation of the ear: variations on an enigma. *In:* Webster D.B., Fay R.R., Popper A.N. (eds.): *The Evolutionary Biology of Hearing.* New York: Springer Verlag, pp. 185-210.

Roberts, B.L., Meredith, G.E., Maslam, S. (1989) Immunocytochemical analysis of the dopamine system in the brain and spinal cord of the European eel, *Anguilla anguilla. Anat. Embryol.* 180: 401-412.

Romeis, B. (1989) *Mikroskopische Technik.* München: Urban & Schwarzenberg.

Ross, L.S., Parrett T., Easter, S.S. jr. (1992) Axonogenesis and morphogenesis in the embryonic zebrafish brain. *J. Neurosci.* 12: 467-482.

Rowe, J.S., Beauchamp, R.D. (1982) Visual responses of nucleus corticalis neurons in the perciform teleost, northern rock bass (*Ambloplites rupestris rupestris*). *Brain Res.* 236: 205-209.

Sakamoto, N., Ito, H. (1982) Fiber connections of the corpus glomerulosum in a teleost, *Navodon modestus. J. Comp. Neurol.* 205: 291-298.

Schellart, N.A.M. (1990) The visual pathways and central non-tectal processing. *In:* Douglas R.H., Djamgoz M.B.A. (eds.): *The Visual System of Fish.* London: Chapman & Hall, pp. 345-372.

Sibbing, F.A., Uribe, R. (1985) Regional specializations in the oropharyngeal wall and food processing in the carp (*Cyprinus carpio* L.). *Neth. J. Zool.* 35: 377-422.

Solnica-Krezel, L., Schier, A.F., Driever, W. (1994) Efficient recovery of ENU induced mutations from the zebrafish germline. *Genetics* 136: 1401-1420.

Song, J., Boord, R.L. (1993) Motor components of the trigeminal nerve and organization of the mandibular arch muscles in vertebrates. *Acta Anat.* 148: 139-149.

Striedter, G.F. (1991) Auditory, electrosensory, and mechanosensory lateral line pathways through the forebrain in channel catfishes. *J. Comp. Neurol.* 312: 311-331.

Striedter, G.F. (1992) Phylogenetic changes in the connections of the lateral preglomerular nucleus in ostariophysan teleosts: a pluralistic view of brain evolution. *Brain Behav. Evol.* 39: 329-357.

Striedter, G.F., Northcutt, R.G. (1989) Two distinct visual pathways through the superficial pretectum in a percomorph fish. *J. Comp. Neurol.* 283: 342-354.

Torres, B., Pastor, A.M., Cabrera, B., Salas, C., Delgado-Garcia, J.M. (1992) Afferents to the oculomotor nucleus in the goldfish (*Carassius auratus*) as revealed by retrograde labeling with horseradish peroxidase. *J. Comp. Neurol.* 324: 449-461.

Uchiyama, H., Matsutani, S., Ito, H. (1988) Pretectum and accessory optic system in the filefish, *Navodon modestus* (Balistidae, Teleostei) with special reference to visual projections to the cerebellum and oculomotor nuclei. *Brain Behav. Evol.* 31: 170-180.

Vanegas, H., Ebbesson S.O.E. (1976) Telencephalic projections in two teleost species. *J. Comp. Neurol.* 165: 181-195.

Villani, L., Dipietrangelo, L., Pallotti, C., Pettazzoni, P., Zironi, I., Guarnieri, T. (1994) Ultrastructural and immunohistochemical study of the telencephalo-habenulo-interpeduncular connections of the goldfish. *Brain Res. Bull.* 34: 1-5.

Webster, D.B., Fay, R.R., Popper, A.N. (1992) *The Evolutionary Biology of Hearing.* New York: Springer Verlag.

Wilkinson, D.G., Krumlauf, R. (1990) Molecular approaches to the segmentation of the hindbrain. *TINS* 13: 335-339.

Wilkinson, D.G., Bhatt, S., Chavrier, P., Bravo, R., Charnay, P. (1989a) Segment specific expression of a zinc finger gene in the developing nervous system of the mouse. *Nature* 337: 461-465.

Wilkinson, D.G., Bhatt, S., Cook, M., Boncinelli, E., Krumlauf, R. (1989b) Segmental expression of *Hox-2* homeobox genes in the developing mouse hindbrain. *Nature* 341: 405-409.

Williams, B., Vanegas, H. (1982) Tectal projections in teleosts: responses of some target nuclei to direct tectal stimulation. *Brain Res.* 242: 3-9.

Wilson, S.W., Easter, S.S. jr. (1991) Stereotyped pathway selection by growth cones of early epiphysial neurons in the embryonic zebrafish. *Development* 112: 723-746.

Wilson, S.W., Placzek, M., Furley, A. (1993) Border disputes: do boundaries play a role in growth-cone guidance? *TINS* 16: 316-323.

Wilson, S.W., Ross, L.S., Parrett, T., Easter, S.S. jr. (1990) The development of a simple scaffold of axon tracts in the brain of the embryonic zebrafish, *Brachydanio rerio*. *Development* 108: 121-145.

Wullimann, M.F. (1988) The tertiary gustatory center in sunfishes is not nucleus glomerulosus. *Neurosci. Lett.* 86: 6-10.

Wullimann, M.F. (1994) The teleostean torus longitudinalis: a short review on its structure, histochemistry, connectivity, possible function, and phylogeny. *Europ. J. Morph.* 32: 235-242.

Wullimann, M.F., Meyer, D.L. (1990) Phylogeny of putative cholinergic visual pathways through the pretectum to the hypothalamus in teleost fish. *Brain Behav. Evol.* 36: 14-29.

Wullimann, M.F., Meyer, D.L. (1993) Possible multiple evolution of indirect telencephalo-cerebellar pathways in teleosts: studies in *Carassius auratus* and *Pantodon buchholzi*. *Cell Tiss. Res.* 274: 447-455.

Wullimann, M.F., Northcutt, R.G. (1988) Connections of the corpus cerebelli in the green sunfish and the common goldfish: a comparison of perciform and cypriniform teleosts. *Brain Behav. Evol.* 32: 293-316.

Wullimann, M.F., Northcutt, R.G. (1989) Afferent connections of the valvula cerebelli in two teleosts, the common goldfish and the green sunfish. *J. Comp. Neurol.* 289: 554-567.

Wullimann, M.F., Northcutt, R.G. (1990) Visual and electrosensory circuits of the diencephalon in mormyrids: an evolutionary perspective. *J. Comp. Neurol.* 297: 537-552.

Wullimann, M.F., Hofmann, M.H., Meyer, D.L. (1991a) The valvula cerebelli of the spiny eel, *Macrognathus aculeatus*, receives primary lateral-line afferents from the rostrum of the upper jaw. *Cell Tiss. Res.* 266: 285-293.

Wullimann, M.F., Meyer, D.L., Northcutt, R.G. (1991b) The visually related posterior pretectal nucleus in the non-percomorph teleost *Osteoglossum bicirrhosum* projects to the hypothalamus: a DiI study. *J. Comp. Neurol.* 312: 415-435.

Yanagihara, D., Watanabe, S., Mitarai, G. (1993a) Neuroanatomical substrate for the dorsal light response I. Differential afferent connections of the lateral lobe of the valvula cerebelli in goldfish (*Carassius auratus*). *Neurosci. Res.* 16: 25-32.

Yanagihara, D., Watanabe, S., Takagi, S., Mitarai, G. (1993b) Neuroanatomical substrate for the dorsal light response II. Effects of kainic acid-induced lesions of the valvula cerebelli on the goldfish dorsal light response. *Neurosci. Res.* 16: 33-37.

Yoshimoto, M., Ito, H. (1993) Cytoarchitecture, fiber connections, and ultrastructure of the nucleus praetectalis superficialis pars magnocellularis (PSm) in carp. *J. Comp. Neurol.* 336: 433-446.

Zottoli, S.J., van Horne, C. (1983) Posterior lateral line afferent and efferent pathways within the central nervous system of the goldfish with special reference to the Mauthner cell. *J. Comp. Neurol.* 219: 100-111.

Printed in the United States
By Bookmasters